T0269513

CAMBRIDGE LIBRARY COLLECTION

Books of enduring scholarly value

Botany and Horticulture

Until the nineteenth century, the investigation of natural phenomena, plants and animals was considered either the preserve of elite scholars or a pastime for the leisured upper classes. As increasing academic rigour and systematisation was brought to the study of 'natural history', its subdisciplines were adopted into university curricula, and learned societies (such as the Royal Horticultural Society, founded in 1804) were established to support research in these areas. A related development was strong enthusiasm for exotic garden plants, which resulted in plant collecting expeditions to every corner of the globe, sometimes with tragic consequences. This series includes accounts of some of those expeditions, detailed reference works on the flora of different regions, and practical advice for amateur and professional gardeners.

The Naturalist's and Traveller's Companion

First published in 1772 and reissued here in its 1799 third edition, this work was intended to provide the traveller with advice on collecting and preserving scientific specimens, and on pursuing intellectual investigations. John Coakley Lettsom (1744–1815) was a physician and philanthropist, and on inheriting his family plantation in 1767, his first action was to free all its slaves. He practised medicine in the West Indies and in London, and wrote on topics which he felt would benefit society. This book is divided into two parts, the first describing methods of forming collections of insects, birds and animals, seeds and plants, and minerals. The second part suggests the sorts of questions and enquiries the traveller should ask about the writings, culture, religion, history and natural history of the lands he is visiting. This offers a fascinating insight into the approach and expectations of the educated traveller in the eighteenth century.

The Naturalist's and Traveller's Companion

John Coakley Lettsom

CAMBRIDGE
UNIVERSITY PRESS

University Printing House, Cambridge, CB2 8BS, United Kingdom

Cambridge University Press is part of the University of Cambridge.
It furthers the University's mission by disseminating knowledge in the pursuit of education, learning and research at the highest international levels of excellence.

www.cambridge.org
Information on this title: www.cambridge.org/9781108076760

This edition first published 1799
This digitally printed version 2015

ISBN 978-1-108-07676-0 Paperback

The original edition of this book contains a number of colour plates, which have been reproduced in black and white. Colour versions of these images can be found online at www.cambridge.org/9781108076760

Selected botanical reference works available in the CAMBRIDGE LIBRARY COLLECTION

al-Shirazi, Noureddeen Mohammed Abdullah (compiler), translated by Francis Gladwin: *Ulfáz Udwiyeh, or the Materia Medica* (1793) [ISBN 9781108056090]

Arber, Agnes: *Herbals: Their Origin and Evolution* (1938) [ISBN 9781108016711]

Arber, Agnes: *Monocotyledons* (1925) [ISBN 9781108013208]

Arber, Agnes: *The Gramineae* (1934) [ISBN 9781108017312]

Arber, Agnes: *Water Plants* (1920) [ISBN 9781108017329]

Bower, F.O.: *The Ferns (Filicales)* (3 vols., 1923–8) [ISBN 9781108013192]

Candolle, Augustin Pyramus de, and Sprengel, Kurt: *Elements of the Philosophy of Plants* (1821) [ISBN 9781108037464]

Cheeseman, Thomas Frederick: *Manual of the New Zealand Flora* (2 vols., 1906) [ISBN 9781108037525]

Cockayne, Leonard: *The Vegetation of New Zealand* (1928) [ISBN 9781108032384]

Cunningham, Robert O.: *Notes on the Natural History of the Strait of Magellan and West Coast of Patagonia* (1871) [ISBN 9781108041850]

Gwynne-Vaughan, Helen: *Fungi* (1922) [ISBN 9781108013215]

Henslow, John Stevens: *A Catalogue of British Plants Arranged According to the Natural System* (1829) [ISBN 9781108061728]

Henslow, John Stevens: *A Dictionary of Botanical Terms* (1856) [ISBN 9781108001311]

Henslow, John Stevens: *Flora of Suffolk* (1860) [ISBN 9781108055673]

Henslow, John Stevens: *The Principles of Descriptive and Physiological Botany* (1835) [ISBN 9781108001861]

Hogg, Robert: *The British Pomology* (1851) [ISBN 9781108039444]

Hooker, Joseph Dalton, and Thomson, Thomas: *Flora Indica* (1855) [ISBN 9781108037495]

Hooker, Joseph Dalton: *Handbook of the New Zealand Flora* (2 vols., 1864–7) [ISBN 9781108030410]

Hooker, William Jackson: *Icones Plantarum* (10 vols., 1837–54) [ISBN 9781108039314]

Hooker, William Jackson: *Kew Gardens* (1858) [ISBN 9781108065450]

Jussieu, Adrien de, edited by J.H. Wilson: *The Elements of Botany* (1849) [ISBN 9781108037310]

Lindley, John: *Flora Medica* (1838) [ISBN 9781108038454]

Müller, Ferdinand von, edited by William Woolls: *Plants of New South Wales* (1885) [ISBN 9781108021050]

Oliver, Daniel: *First Book of Indian Botany* (1869) [ISBN 9781108055628]

Pearson, H.H.W., edited by A.C. Seward: *Gnetales* (1929) [ISBN 9781108013987]

Perring, Franklyn Hugh et al.: *A Flora of Cambridgeshire* (1964) [ISBN 9781108002400]

Sachs, Julius, edited and translated by Alfred Bennett, assisted by W.T. Thiselton Dyer: *A Text-Book of Botany* (1875) [ISBN 9781108038324]

Seward, A.C.: *Fossil Plants* (4 vols., 1898–1919) [ISBN 9781108015998]

Tansley, A.G.: *Types of British Vegetation* (1911) [ISBN 9781108045063]

Traill, Catherine Parr Strickland, illustrated by Agnes FitzGibbon Chamberlin: *Studies of Plant Life in Canada* (1885) [ISBN 9781108033756]

Tristram, Henry Baker: *The Fauna and Flora of Palestine* (1884) [ISBN 9781108042048]

Vogel, Theodore, edited by William Jackson Hooker: *Niger Flora* (1849) [ISBN 9781108030380]

West, G.S.: *Algae* (1916) [ISBN 9781108013222]

Woods, Joseph: *The Tourist's Flora* (1850) [ISBN 9781108062466]

For a complete list of titles in the Cambridge Library Collection please visit:
www.cambridge.org/features/CambridgeLibraryCollection/books.htm

THE

NATURALIST's AND TRAVELLER's

COMPANION.

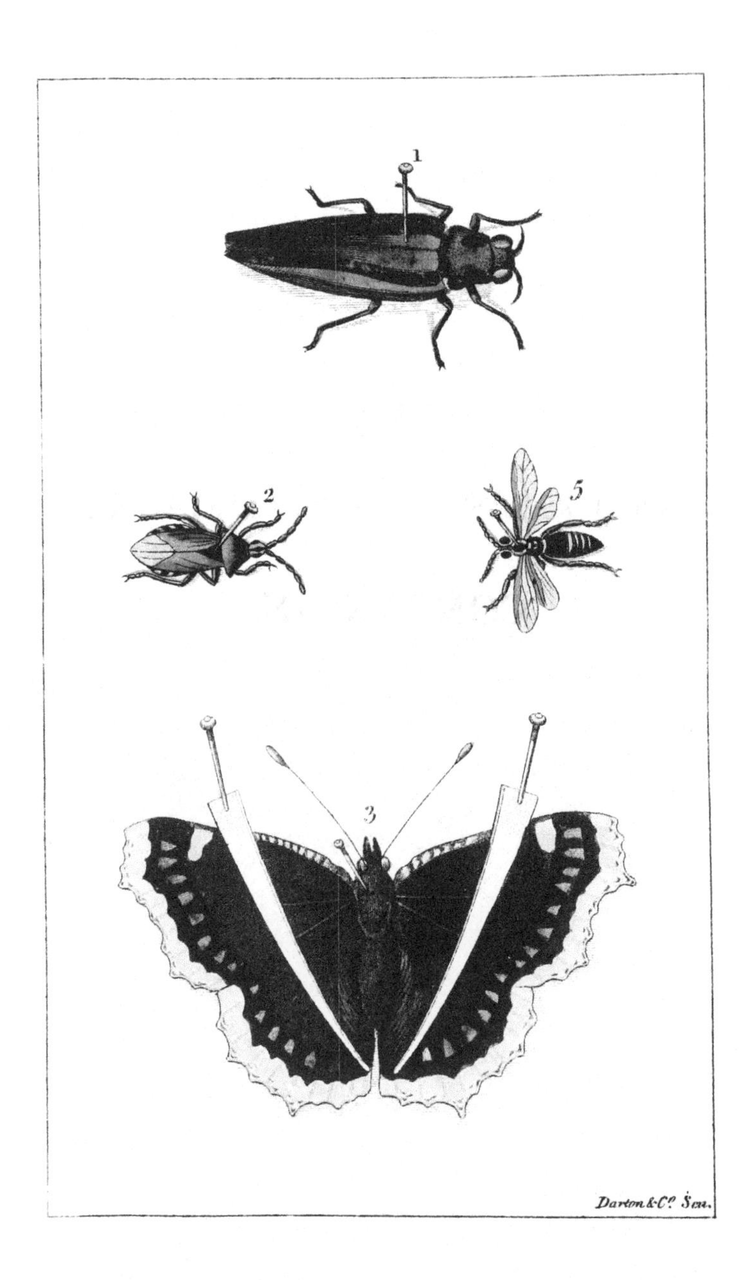
1
2
5
3
Darton & Co. Sc.

THE

NATURALIST'S

AND

TRAVELLER'S

COMPANION;

BY

JOHN COAKLEY LETTSOM. M.D.

the Third Edition.

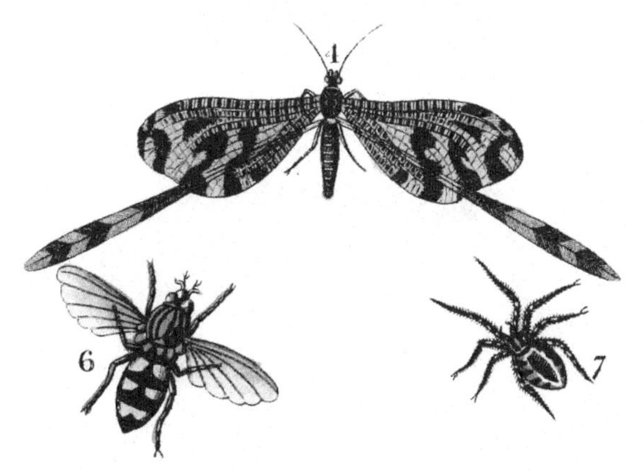

London, C. Dilly. 1799.

CONTENTS.

PART the FIRST.

SECT. VI.

PART the SECOND.

SECT. I.

SECT. II.

SECT. III.

SECT. IV.

SECT. V.

SECT. VI.

SECT. VII.

PRE-

PREFACE.

"HE that enlarges his curiofity after the works of nature," fays a celebrated writer, "demonftrably multiplies the inlets to happinefs. A man that has formed a habit of turning every new object to his entertainment, finds in thefe productions an inexhauftible ftock of materials upon which he can employ himfelf, without any temptations to envy or malevolence; faults, perhaps, feldom totally avoided by thofe, whofe judgment is much exercifed upon the works of art. He has always a certain profpect of difcovering new reafons for adoring the fovereign Author of the univerfe, and probable hopes of making fome difcovery of benefit to others, or of profit to himfelf."

No method appears better calculated to enlarge our knowledge of Natural Hiftory, than vifiting foreign countries, and carefully attending

attending to the different objects they afford, which more or less delight by their novelty and variety; but our inquiries should not be confined merely to private gratification; there are duties of a more rational nature; to be useful to society by distributing happiness amongst our fellow creatures, is one of the highest and most necessary. The numerous products of nature, their application to the wants, the comforts, and even ornaments of life; the manners, customs, and opinions of mankind; agriculture, manufactures, and commerce; the state of arts, learning, and the laws of different nations, when judiciously investigated, tend to enlarge the human understanding, and to render individuals wiser, better, and happier.

The introduction of the common potatoe, the management of silk-worms, the discovery of jesuits bark, the uses of cochineal, lacca and indigo, are undeniable proofs of the advantages which might be derived from the inquiries of ingenious men. The discovery of another such root as the potatoe, another such article of commerce and apparel as silk, another such remedy as the bark, and such other dying articles as cochineal and indigo, would prove acquisitions of the greatest importance to a trading

ing nation, and render the inquiſitive traveller conſpicuous as a public bleſſing.

Many gentlemen and ſea-faring perſons who go abroad, by their office and ſituation in life, enjoy both time and opportunity for collecting the beſt information on ſuch ſubjects of general utility, eſpecially the natural productions peculiar to the place they viſit or reſide in, which they are induced to overlook, for want of proper directions for diſtinguiſhing and preſerving them, whereby things of value and uſe are loſt to the public, and the time of the traveller leſs beneficially employed.

To promote an application of the time and talents of ſuch perſons to rational and commendable inquiries of this kind, is the deſign of the following directions, which the author thinks himſelf juſtified in recommending, as they principally reſult from experiment and obſervation: Theſe were firſt printed in the year 1772, and the reception from the publick was ſuch as to encourage a ſecond edition in 1774; but this likewiſe being ſoon out of print, a third edition was not long afterwards prepared for the preſs, and ſome of the firſt ſections were printed off ſeveral years ago;

various avocations then intervening, prevented the completion of the remaining sections; nor would this little performance have been now resumed, had any publication appeared calculated to preclude the utility of the original plan, which has since been considerably improved, to make it more deserving of future encouragement.

In the second part are introduced several queries and observations on natural history, and upon subjects in general, which have not been clearly and sufficiently ascertained, and therefore merit the attention of the naturalist and traveller.

The principal of those writers who have appeared in different departments of Natural History, will be noticed under the particular heads they treat upon.

If persons who go abroad, or reside in foreign countries, were acquainted with mathematics and drawing, they would in all probability make their remarks more acceptable, by adding accurate maps of the countries they visit or reside in; and by joining to them the drawings of men, their dresses, utensils, weapons, coins, machines, rites, sacrifices, buildings, temples, idols, and

and antiquities; as well as the curious quadrupeds, birds, reptiles, fiſh, inſects, and ſhells peculiar to each place; with the plants found in thoſe climates, eſpecially ſuch as are employed for food, in commerce, manufactures, phyſic, dying and other purpoſes.

In the drawings and deſcriptions relative to natural hiſtory, it is neceſſary to attend to many circumſtances which are the characteriſtics of each ſpecies of the animal and vegetable creation. In quadrupeds the number and diſpoſition of the teeth; ſhape and poſition of the horns; number of the toes in each foot; ſhape and ſize of the claws and hoofs; ſize of the ears; colour and diſpoſition of the whiſkers; nature and growth of the hair in the fur, mane, and tail; length and uſes of the tail; whether calculated to graſp any object, or to give the animal ſtability; and even the attitude which is characteriſtic of the animal, and ſhews beſt it's marks, ſpots, ſtripes, claws, ears, tail, &c. ought to be expreſſed.

In birds, the ſhape and uſes of the bill, whether notched, ſerrated, or otherwiſe remarkable; number and diſpoſition of the toes, and whether diſtinct, lobated, or palmated; length of the legs and nakedneſs of the knees; colour of the greater and ſmaller quill feathers, upper and under coverts of the wings; number and colour of the tail feathers, and coverts of the tail; appearance of the vent, belly, breaſt, throat, back, creſt, wattles, carunculæ, ſpurs, &c.; attitude peculiar to the bird, and the difference between males and females, and young and old birds, ſhould be deſcribed.

In tortoiſes and turtles the diſpoſition of the ſhell, and it's compartments; ſhape and number of it's toes; abſence or preſence of the tail, and ſhape of the head, muſt be delineated.

In ſnakes, the ſcales above and below, their number, colour, and figure; form of their heads, and whether they are venomous or not, ſhould be remarked.

In fiſh, it is neceſſary to attend to the proportion of the breadth to the length; form of the head and diſpoſition of the palate and teeth; ſhape and poſition of the mouth; ſize and ſituation of the eyes; coverts of the gill, and rays of it's underpart; ſpines, horns, and protuberances of the head; number, figure, ſize, and colour

of

of the fins and tail, with the ſpinoſe and ſoft rays in each; the turn of the lateral line, with the form, colour and diſpoſition of the ſcales.

In inſects, the ſeaſon when each of the different kinds appear ſhould be obſerved; the number, ſubſtance, and particular ſhape of their wings, with the poſition of them when the inſect is at reſt; the ſhape of the antennæ or horns, with the number of joints in each; the form of the head, mouth and eyes, more particularly of the head in beetles, of the mouth in bees, waſps, flies, and gnats, and of the eyes in ſpiders; the number and ſize of the legs; the ſhape of the thighs, feet and claws; the ſtings peculiar to the hymenoptera claſs, and the uſes they are applied to: but the natural hiſtory of inſects ſhould in a peculiar manner engage the traveller's attention, as it is of more conſequence to diſcover the natural hiſtory of one deſtructive or uſeful inſect, than merely to collect and bring over twenty in their perfect ſtate; the former, at the ſame time that it makes the ſcience more entertaining, bids fair to benefit mankind, while the latter ſerves only to fill the cabinets of the curious; he ſhould therefore carefully obſerve the manner in which inſects

ſects copulate, and the places where they depoſit their eggs; what food the young larvæ or caterpillars feed upon; if vegetable, whether it be the root, trunk, leaf, flower or fruit; if deſtructive, as they moſtly are, the methods uſed by the natives to deſtroy them; and if uſeful, the means of cultivating them; and what are their natural enemies; the form, attitude, and markings of the caterpillar ſhould be deſcribed; if it has feet, their number, and the particular rings on which they are ſituated; whether it be ſmooth, hairy or ſpinous, and the manner of it's changing into the chryſalis or pupa ſtate, and how long it continues before it arrives to perfection, with the various inſtincts and contrivances they have for avoiding dangers and catching their prey.

In ſhells, not to neglect the number of them belonging to one animal; when ſingle, the turn in the windings, whether to the right or left; the ſtripes, ſpots, bands, knobs, ſpines, furrows and other marks; ſhape of the mouth or opening of the ſhell, lips and beaks: In thoſe that have two or more valves, their equal or different ſizes; the form of the hinges where they are connected, and the number of indentures tallying together; the ſtripes and furrows on the

the outſides, and whether longitudinal or tranſverſal; and the animals inhabiting the ſhells ſhould likewiſe be obſerved and delineated.

In the reſt of the worm tribe, the ſhape, arms, and other parts of the animal ſhould be delineated.

In plants the greateſt accuracy is requiſite, the ſhape of the flower being ſo varied, nice attention is neceſſary to diſtinguiſh it's minute parts; the figure and number of the flower leaves; the form and ſections of the flower-cup; the number and diſpoſition of the duſt veſſels, and of the columnar veſſels ſtanding on the fructification (which are reckoned by botaniſts to be the male and female parts of the flower, and in ſome inſtances are on different plants; in others on the ſame plant, but in different parts of it); the ſhape, ſtructure, and colour of the ſtalk and leaves; the appearance and ſtructure of the roots, and ſuch other circumſtances as characterize the different ſpecies of plants, ought never to be omitted.

The following method of preſerving ſeeds, by John Sneyd, eſq. ſhould have been inſerted, under the directions for bringing over ſeeds, &c. § III. p. 23, which I have extracted from the Tranſactions of the Society for the Encouragement of Arts, Vol. xvi. p. 265. It is merely packing up ſeeds in abſorbent paper, and ſurrounding the ſame by raiſins, or brown moiſt ſugar; which, by experiment, ſeems to afford that genial moiſture requiſite to preſerve the ſeeds in a ſtate fit for vegetation.

The Naturaliſt ſhould endeavour to keep an accurate journal, wherein all the occurrences, obſervations, places, diſtances, deſcriptions, accounts, informations, and remarks, ſhould regularly and daily be entered, while recent in memory.

It would be adviſable to write on a label fixed to each object, a number correſponding to the notes in the journal; by which means, at any future period, the object itſelf may be clearly aſcertained.

THE

NATURALIST's and TRAVELLER's

COMPANION, &c.

PART the FIRST.

SECT. I.

The Method of catching and preserving Insects *for Collections.*

——— Ten thousand different tribes
People the blaze. To sunny waters some
By fatal instinct fly.
———Through the green-wood glade
Some love to stray; there lodg'd, amus'd and fed,
In the fresh leaf. Luxurious, others make
The meads their choice, and visit every flower,
And every latent herb. (*a*)

INSECTS in general are known to most people, the systematic distinctions but to few; nor have we any English names for the greatest part of them. The general denomination of beetles, butterflies, moths, flies, bees, wasps,

(*a*) Thomson's Seasons, Summer, l. 246.

and

and a few other common names, are all that our language ſupplies. It would, therefore, be in vain to enumerate the immenſe variety of genera and ſpecies to any perſon unſkilled in the ſcience of entomology: we may, however, give directions under general names, where to find and how to catch each kind. (*b*)

I. The Coleoptera (*c*), or firſt great claſs of inſects, including beetles, are found in and under the dung (*d*) of animals, eſpecially of cows, horſes, and ſheep: many of them make holes under the dung three or four inches deep; it will therefore be neceſſary to have an iron ſpade to dig them out, when in ſearch of this tribe of inſects.

Some (*e*) are found in rotten and half decayed wood, and under the decayed bark of trees; on the carcaſes (*f*) of animals that have been dead four or five days; on moiſt bones that have been gnawed by dogs or other animals; on flowers having a fœtid ſmell; and on ſeveral kinds of fungous ſubſtances, particularly the

(*b*) Vide Schoeffer. Elementa Entomologica. Curtis's accurate inſtructions for collecting and preſerving inſects, and his introduction to the knowledge of inſects tranſlated from the Fundamenta Entomologiæ of Linnæus. Amæn. Acad. v. 7.

(*c*) Coleoptera, from κολεος, a ſheath, and πτερον, a wing, are ſuch inſects as have cruſtaceous Elytra, or ſhells, which ſhut together, and, form a longitudinal ſuture down to the back of the inſect, as the beetle, Bupreſtris ignita, fig. 1.

(*d*) Scarabæus, *chafer*. Dermeſtes, *leather-eater*. Hiſter, *mimick-beetle*. Staphylinus, *rove-beetle*. (*e*) Lucanus, *ſtag-beetle*. Cerambyx, *capricorn-beetle*. (*f*) Hiſter. Silpha, *carrion-beetle*.

(*g*) Byrrhus,

the rotten and moſt ſtinking: others (*g*) may be found in a morning about the bottoms of perpendicular rocks and ſand banks, and alſo upon the flowers of trees and herbaceous plants.

Many kinds (*h*) may be caught in rivers, lakes, and ſtanding pools, by means of a thread net, with ſmall meſhes, on a round wire hoop, fixed at the end of a long pole.

In the middle of the day, when the ſun ſhines hot, (*i*) ſome are to be ſeen on plants and flowers, blighted trees and ſhrubs; others (*k*) in moiſt meadows are beſt diſcovered at night, by the ſhining light which they emit.

A great variety (*l*) ſit cloſe on the leaves of plants, particularly of the burdock, elecampane, coltsfoot, dock, thiſtle, and the like; or feed on different kinds of tender herbs (*m*).

Numbers (*n*) may be found in houſes, dark cellars, damp pits, caves, and ſubterraneous paſſages, or on umbelliferous flowers (*o*), on the trunks as well as the leaves of trees; in timber-yards, and in the holes of decayed wood.

Some (*p*) inhabit wild commons, the margins of pools, marſhes, and rivulets; and are likewiſe ſeen creeping on flags, reeds, and all kinds of water-plants.

Multitudes (*q*) live under ſtones, moſs, rubbiſh, and wrecks near the ſhores of lakes and

(*g*) Byrrhus, curculio, *weevil.* Bruchus, *ſeed-beetle.* (*h*) Gyrinus, *whirl-beetle.* Dytiſcus, *water-beetle.* (*i*) Coccinella, *lady-fly.* Bupreſtis, *burn-cow.* Chryſomela, *golden honey-beetle.* Cantharis, *ſoft-winged beetle.* Elater, *ſpring-beetle.* Necydalis, *clipt-winged beetle.* (*k*) Lampyris, *glow-worm.* (*l*) Caſſidë, *tortoiſe-beetle.* (*m*) Meloa, *bliſter-beetle.* (*n*) Tenebrio, *ſtinking-beetle.* (*o*) Cerambyx, Ptinus. (*p*) Leptura, *wood-beetle.* Cicindela, *gloſſy beetle.* (*q*) Carabus, *ground-beetle.*

rivers.

rivers. Theſe are found alſo in bogs, marſhes, moiſt places, pits, and holes of the earth, on ſtems of trees; and in an evening they crawl plentifully along path-ways after a ſhower of rain.

Some (*r*) may be diſcovered in the hollow ſtems of decayed umbelliferous plants, and on many ſorts of flowers and fruits.

II. Another claſs (*s*) of inſects are found about (*t*) bake-houſes, corn-mills, in ſhips, and in all places where meal is kept; on graſs (*u*), and all kinds of field herbage. Some (*v*) of theſe frequent rivers, lakes and ſtanding pools.

III. Butterflies and moths make another great diviſion (*w*). In the day, when the ſun is warm, butterflies (*x*) are ſeen on many ſorts of trees, ſhrubs, plants, and flowers. Moths (*y*) may be ſeen in the day-time, ſitting on walls, pales, trunks of trees, in ſhades, out-houſes, dry holes, and crevices *; on fine evenings they fly about the

(*r*) Forficula, *earwig*.

(*s*) Hemiptera, from ημισυ, half, and πτερον, a wing, have their upper wings uſually half cruſtaceous, and half membranaceous, not divided by a longitudinal ſuture, but incumbent on each other, as the Cimex, fig. 2. (*t*) Blatta, *cockroach*, (*u*) Mantis, *camel cricket*. Gryllus, *locuſt*. Fulgora, cicada, *flea-locuſt*. Cimex, *bug*. (*v*) Notonecta, *boat-fly*. Nepa, *water-ſcorpion*.

(*w*) Lepidoptera, from λεπις, a ſcale, and πτερον, a wing, are inſects having four wings, covered with fine ſcales in the form of powder or meal, as in the butterfly, Papilio Antiopa, fig. 3. (*x*) Papilio. *butterfly*. (*y*) Phalæna, *moth*.

* The beſt method to kill moths, when ſtanding, or ſitting on walls

the places they inhabit in the day time. Some (*z*) are ſeen flying in the day-time over the flowers of honey-ſuckles and other plants with tubular flowers. Inſects of this ſpecies ſeldom ſit to feed, but continue vibrating on the wing, while they thruſt the tongue or proboſcis into the flowers.

IV. Inſects of this claſs (*a*) are found in woods (*b*), hedges, meadows, ſand-banks, walls, pales, fruits, and unbelliferous flowers; ſome (*c*) fly about lakes and rivers in the day.

V. The fifth diviſion (*d*) including waſps (*e*), bees, (*f*), &c. may be ſeen about hedges (*g*), ſhrubs, flowers, and fruits. Waſps and bees are the only winged inſects that have any great degree of poiſon in them; they ſhould therefore be taken with a pair of forceps, and handled cautiouſly on account of their ſtings, which are dangerous

walls, pales, &c. is to give them a gentle preſſure on the thorax, or breaſt, with the end of any thing ſmooth, as a tobacco-ſtopper, or tooth-pick caſe: this will deprive them of ſenſe and motion for the preſent moment, and they will fall down on whatever is held under them; a pin ſhould then be ſtruck through the thorax, which ſhould be preſſed between the thumb and finger of the left-hand, till it is felt to give way a little; they may then be ſtuck in the receiving box.

(*z*) Spinx, *hawk-moth*.

(*a*) Neuroptera, from νευρον, a nerve, and πτερον, a wing, have four membranaceous tranſparent naked wings, generally like network, as in the Panorpa Coa, fig. 4. (*b*) Myrmeleon, hemerobius, *pearl-fly*. Raphidia, *camel-fly*. (*c*) Libellula, *dragon-fly*, Ephemera, *may-fly*. Phryganea, *ſpring-fly*.

(*d*) Hymenoptera, from υμην, a membrane, and πτερον, a wing. Inſects with four membranaceous wings, tail furniſhed with a ſting; as in the Tenthredo, fig. 5. (*e*) Veſpa, *waſp*. (*f*) Apis, *bee*. (*g*) Tenthredo, *ſaw-fly*. Sirex, *tailed waſp*. Ichneumon,

dangerous. Some (*h*) of this diviſion have ſtings, but no poiſon, and are to be found on the flowers of umbelliferous plants, when the ſun ſhines hot in the middle of the day; at which time others (*i*) are ſeen on ſand-banks, walls, and pales.

VI. Flies of various kinds conſtitute the next claſs (*k*); they fly about the tops of (*l*) trees, little hills, horſes, cows, ſheep, ditches, dung-hills, and every offenſive object. Some (*m*) are found on all ſorts of flowers, particularly thoſe of a fœtid ſmell. Many (*n*) of theſe are moſt eaſily taken when they begin to feed; for in the middle of the day they are ſo quick and active, that it is almoſt impoſſible to catch them.

VII. The laſt great diviſion (*o*) contains ſcorpions, ſpiders, crabs, lobſters, &c. It is neceſſary only to obſerve here, that all kinds of inſects having no wings, may be preſerved in ſpirits, brandy or rum, except crabs, lobſters, and the like, which may conveniently be preſerved dry.

Under each claſs of inſects, I ſhall relate the

mon, *ichneumon-fly*. Sphex, *ichneumon-waſp*. Veſpa. Apis. (*h*) Mutilla, *naked-bee*. (*i*) Chryſis.

(*k*) Diptera, from δυω two, and πτερον, a wing, are ſuch as have only two wings, and poiſers, as in the fly, fig. 6. (*l*) Oeſtrus, *gad-fly*. Muſca, *fly*. Tabanus, *whame*. Hippoboſca, *horſe-fly*.

(*m*) Tipula. Conops. Aſilus, *waſp-fly*. (*n*) Bombylius, *flower-breeze*.

(*o*) APTERA, from α, without, and πτερον, a wing, inſects having no wings, as the ſpider, fig. 7.

methods

1

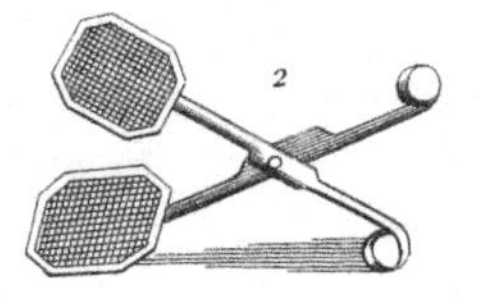
2

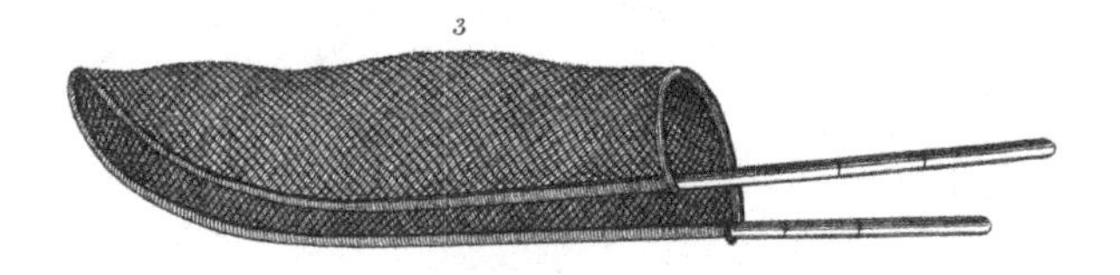
3

methods of killing them the moſt readily, and with the leaſt pain; as the purſuit of this part of natural hiſtory hath often been branded with cruelty; and however reaſonably the naturaliſt may exculpate himſelf by pleading the propriety of ſubmitting to an evil, which leads to uſeful diſcoveries, yet for wanton cruelty there never can be a juſt pretext.

> —The poor beetle that we tread upon,
> In corp'ral ſufferance finds a pang as great
> As when a giant dies (*p*).

I. The firſt claſs, conſiſting of beetle (coleoptera) are hard-winged. Many kinds fly about in the day, others in the evening, ſome at night only. They may be caught with a gauze net, (pl. 2, fig. 2.) or a pair of forceps covered with gauze (pl. 2, fig. 2). When they are taken, ſtick a pin through the middle of one of the hard wings, and paſs it through the body, as in pl. 1. fig. 1, frontiſpiece. They may be killed inſtantly, by immerſion in hot water, as well as in ſpirit of wine; then ſtick them on a piece of cork, and afterwards carefully place their legs in a creeping poſition, and let them continue expoſed to the air until all the moiſture is evaporated from their bodies. Beetles may alſo be preſerved in ſpirit of wine, brandy or rum, cloſely corked up.

II. Inſects of the ſecond claſs (hemiptera) may be killed in the ſame manner as beetles, and likewiſe by means of a drop of the etherial oil of turpentine applied to the head; or in the manner

(*p*) Shakeſpeare's Meaſure for Meaſure.

manner to be defcribed under the next clafs for killing moths.

III. The divifion of butterflies and moths (lepidoptera) as well as all flies with thin membranaceous wings, fhould be catched with a gauze net, or a pair of gauze forceps: (see pl. 1, fig. 2) when taken in the forceps, run a pin through the thorax or fhoulders, between the fore-wings, as in frontifpiece, fig. 3. After this is done, take the pin by the head, and remove the forceps, and with the other hand pinch the breaft of the infect, in order to deprive it of fenfation and life: the wings of butterflies fhould be expanded, and kept fo, by the preffure of fmall flips of paper, for a day or two. Moths expand their wings when at reft, and they will naturally take that pofition.

The larger kinds of thefe infects will not fo readily expire by this method, as by fticking them upon the bottom of a cork exactly fitted to the mouth of a bottle, into which a little fulphur had been put, and by gradually heating the bottle till an exhalation of the fulphur takes place, when the infect inftantly dies, without injuring its colours or plumage.

Perhaps the moft eafy method of killing moths and butterflies is by means of a needle made of fteel fixed in an ivory handle, which muft have paffed through the body of the moth; the thumb and forefinger of the left-hand muft be inftantly placed on each fide of the cheft of the moth, juft under the wings, when a fqueeze will entirely deprive the infect of motion, and probably of fenfation too for fome minutes; thefe, however, will return

turn unlefs prevented by the following method.

The point of the needle muft be paffed through a fmall hole in thin brafs, or tin plate, and then held in the flame of a candle for about half a minute, which will effectually deftroy the life of the infect.

Although the idea of burning the infect to death may appear cruel at firft fight, yet it fhould be remembered, that this is not done till after the infect has been deprived of fenfation, by preffure firft recommended; fo that, on the whole, it appears to be the moft eafy and, fpeedy method of deftroying them hitherto practifed.

It is hardly neceffary to add, that a brafs or tin plate, for which a piece of card or pafteboard may be fubftituted, is ufed to prevent the moth from receiving any injury from the candle.

The beft method of having the moft perfect butterflies, is to find out, if poffible, the larva or caterpillar of each, by examining the plants, fhrubs or trees they ufually feed upon, or by beating the fhrubs and trees with long poles, and thereby fhaking the caterpillars into a fheet fpread underneath to receive them; to put them into boxes covered with thin canvas, gauze, or cat-gut (pl. 2, fig. 1) and to feed them with the frefh leaves of the tree or herb on which they are found; when they are full-grown, they will change into the pupa, or chryfalis ftate, and require then no other care, till they come out perfect butterflies, at which time they may be killed, as before directed. Sometimes thefe inſects may be found hanging on walls, pales, and branches of trees, in the chryſalis ſtate.

Moths might likewiſe be procured more perfect, by collecting the caterpillars, and breeding them in the ſame manner as butterflies. As the larvæ or caterpillars cannot be preſerved dry, nor very well kept in ſpirit, it would be ſatisfactory if exact drawings could be made of them while they are alive and perfect. It may be neceſſary to obſerve, that in breeding theſe kinds of inſects, ſome earth ſhould be put into the boxes, as likewiſe ſome rotten wood in the corners, and ſome moſs ſhould be put upon it, which ſhould always be kept damp; becauſe, when the caterpillars change into the pupa, or chryſalis ſtate, ſome go into the earth, and continue under ground for many months before they come out into the moth ſtate; and ſome cover themſelves with a hard ſhell, made up of ſmall pieces of rotten wood. Hence alſo, as many go into the earth, valuable inſects may ſometimes be found by digging after them a foot deep, about the roots of trees, ſhrubs, and plants.

IV. The fourth claſs of inſects (neuroptera) may be killed with ſpirit of wine, oil of turpentine, or by the fumes of ſulphur.

V. Thoſe of the next claſs (hymenoptera) may be killed in the ſame manner. A pin may be run through one of their wing-ſhells and body, as repreſented in pl. 1, fig. 5.

VI. Inſects of the ſixth claſs (diptera) may likewiſe

likewise be killed by spirit, or by fumes of sulphur.

VII. Those of the last division (aptera) are in general, subjects which should be kept in spirit.

When in search of insects, we should have a box suitable to carry in the pocket, lined with cork at the bottom and top to stick them upon, until they are brought home. If this box be strongly impregnated with camphor, the insects soon become stupefied, and are thereby prevented from fluttering and injuring their plumage. Besides a gauze forceps (pl. II, fig. 7) the collector should have a large musqueto gauze net, made in the shape of a bat fowling-net, (pl. II, fig. 3) and also a pin-cushion with three or four different sizes of pins to suit the different sizes of insects.

In hot climates, insects of every kind, but particularly the larger, are liable to be eaten by ants and other small insects, especially before they are perfectly dry; to avoid this, the piece of cork on which our insects are stuck in order to be dried, should be suspended from the cieling of the room, by means of a slender string or thread; besmear this thread with bird-lime, or some adhesive substance, to intercept the rapacious vermin of these climes in their passage along the thread.

After our insects are properly dried, they may be placed in the cabinet or boxes where they are to remain: these boxes should be kept dry; and also made to shut very close, to prevent small insects from destroying them; the bottoms of the boxes should be covered with pitch, or green wax, over which paper may be laid; or, which is better lined with cork, well impregnated with a ſolution of corroſive ſublimate mercury.

The fineſt collections have been ruined by ſmall inſects, and it is impoſſible to have our cabinets too ſecure. Such inſects as are thus attacked may be fumigated with ſulphur, in the manner deſcribed for killing moths; if this prove ineffectual, they may be immerſed in ſpirit of wine, without much injuring their fine plumage or colors; and afterwards let them be ſprinkled about their bodies and inſertions of the wings with the ſolution above-mentioned. But baking the inſects in an oven in the manner to be deſcribed in the next ſection for birds, is the moſt effectual method of extirpating theſe enemies; however the utmoſt caution is requiſite in this proceſs, in regulating the heat of the oven.

Theſe obſervations and directions reſpecting inſects, may, perhaps, be the means of exciting the curioſity of ſome, whoſe inquiries after this part of natural hiſtory will be amply compenſated by the frequent opportunities of enlarging their knowledge, as there is ſcarce any part of the ſurface of this globe, ſcarce a tree, a ſhrub, or a plant, an animal either living or dead, or even the excrements of animals, on which ſome kind of inſect does not depend for its ſubſiſtence and propagation.

——— The flowery leaf
Wants not its soft inhabitants. Secure,
Within its winding citadel, the stone
Holds multitudes. But chief the forest-boughs,
That dance unnumber'd to the playful breeze,
The downy orchard, and the melting pulp
Of mellow fruit, the nameless nations feed
Of evanescent insects. Where the pool
Stands mantled o'er with green, invisible,
Amidst the floating verdure, millions stray.
Each liquid too, whether it pierces, soothes,
Inflames, refreshes, or exalts the taste,
With various forms abounds. Nor is the stream
Of purest crystal, nor the lucid air,
Though one transparent vacancy it seems,
Void of their unseen people——(*q*).

(*q*) Thomson's Seasons, SUMMER, l. 296.

SECT.

SECT. II.

Method of preserving BIRDS *and other Animals,*

——— Vitam excoluere per Artes.
VIR. Æn. 6. v. 663.

THE general increase of knowledge of late in natural history, from the attention of individuals to various branches of it, must afford no small degree of pleasure to the sensible part of mankind. Whilst such different researches have given entertainment to different dispositions, enlarged the mind, and engaged and diverted it from unprofitable or dangerous pursuits, they have occasionally given rise to the most useful improvements in every department of life, and afforded means of joining utility with elegance.

To promote these purposes more effectually, a more general knowledge of a good antiseptic for animal substances has been much enquired after. Owing to a want of this, many curious animals, and birds particularly, come to our hands in a very imperfect state: some from foreign parts entirely miscarry, and others of the finest plumage are devoured by insects.

To

To promote these purposes more effectually, a knowledge of the means of preserving birds and other animals must be highly desirable. The methods made use of, by captain Davies, and T. S. Kuckahn, have been published in the philosophical Transactions (*r*).

The former directs birds in perfect plumage, " to be opened from the upper part of the breast " to the vent, with a sharp knife, or pair of scissars, " the feathers of the breast and belly being first " carefully laid aside by the fingers, so as not to " hinder the skin being easily come at. The " skin must then be carefully loosened from all " the fleshy parts of the breast, body, thighs, " and wings; after this, cut off all the flesh from " those parts, and take out also the entrails and " all the inside; then, having got a composition " of burnt alum, camphor, and cinnamon, of each " an equal quantity, well powdered and mixed " together; strew some of this powder lightly " over the whole carcase; but salt is by no " means to be used in this composition, as it al- " ways will drop and nasty the plumage in moist " weather: pour also into the body a small " quantity of camphor dissolved in rectified spi- " rits of wine; after that, fill up the cavity with " fine cotton, or any soft woolly substance, pour- " ing some of the aforesaid spirits into the cot- " ton or stuffing. Open next the mouth, and " with a pair of scissars take away the tongue, " the roof of the mouth, eyes, brains, and in- " side of the head; fill that also with the same " composition; and, having procured eyes as " near the natural ones as possible, put them

(*r*) Vol. IX. anno 1770, p. 184, and 302.

into

" into the ſockets by means of a ſmall pair of
" nippers introduced at the mouth. The eyes
" will be beſt made by letting fall ſome drops
" of black ſealing wax on a card of the ſize of
" the natural ones (s); the card muſt be cut
" ſomething larger than the wax, to prevent
" their falling out of the head. Fill the head
" quite full with cotton, pouring ſome of the
" ſpirits down the throat, with ſome of the pow-
" der; a ſmall piece of braſs wire, that has been
" heated in the fire to make it pliable, may be
" put down the throat, being paſſed through one
" of the noſtrils, and faſtened to the breaſt bone,
" to place the head in any attitude you chooſe;
" next fill up the body, where the fleſh has been
" taken away, with cotton and your compoſition;
" and having a fine needle and ſilk, ſew up the
" ſkin, beginning at the breaſt, obſerving, as you
" approach towards the vent, to ſtuff the ſkin as
" tight as it will bear. This will be eaſieſt ac-
" compliſhed by means of a ſmall piece of ſtick
" or ivory, like a ſkewer, till the whole is done:
" then lay your feathers of the breaſt and belly in
" their proper order, and your bird will be com-
" pleted. If you would chuſe to put it into an
" attitude, by introducing a ſmall piece of the
" wire above-mentioned through the ſole of
" each foot up the leg, and into the pinion of

(s) Wax is not a proper ſubſtance for eyes; there are perſons in London, whoſe buſineſs it is to make glaſs eyes of any ſize or colour, at a trifling expence, but as theſe cannot be had in voyages, it would be proper, while the bird is freſh, to take a drawing of the eyes in colours.

" each wing (*u*) it may be diſpoſed of as you
" pleaſe."

Inſtead of uſing the ſolution of camphor in ſpirit of wine, Kuckahn reccommends a liquid varniſh, made of two pounds of common or raw turpentine, one pound of camphor, and one pound of ſpirit of turpentine. The camphor is to be broke into very ſmall pieces, and the whole is to be put into a glaſs veſſel, open at top, which is to be placed in a ſand heat, till the ingredients are perfectly diſſolved.

For the dry compound of cinnamon, burnt alum, and camphor, directed in the foregoing method, he ſubſtitutes the two following compoſitions.

Corroſive ſublimate, — —	¼ lb.
Saltpetre prepared or burnt, —	½ lb.
Alum burnt, — —	¼ lb.
Flowers of ſulphur, — —	½ lb.
Muſk (*x*), — — —	¼ lb.
Black pepper, — —	1 lb.
Tobacco ground coarſe. —	1 lb.

(*u*) The late Mr. Leman, who was remarkable for the eaſy attitude of his birds, paſſed a wire ſideways through one wing into the breaſt-bone, the other end of the wire being faſtened into the box incloſing the bird.

(*x*) The muſk renders this compoſition very expenſive for which the ſame quantity of camphor or myrrh might be ſubſtituted.

Mix the whole together, and keep it in a glaſs veſſel ſtopped cloſe. Some of this is to be ſtrewed upon the inſide of the ſkin and cavity of the head, after they have been waſhed with the varniſh.

The other dry compoſition (*y*) is made of equal quantities of tanſy, wormwood, hops, and tobacco, which are to be cut ſmall and mixed together; with this the cavities of the craw and body are to be ſtuffed. He likewiſe directs an artificial breaſt to be made of ſoft wood, and fitted to the proper place, after being moiſtened with the varniſh (*z*).

I thought it might be acceptable to lay before the reader, the above methods, which were practiſed ſome years ſince in this country; but at preſent, a very ſimple one is purſued, which is at the ſame time preferable.

In preſerving all animals, the principal objects ſhould be to remove not only the fleſh of the ſubject but likewiſe as much as poſſible, the bones, which are equally liable to putreſcency, and to invite inſects. In ſhort, after opening the bird,

(*y*) This is entirely uſeleſs, and forms a leſs ſoft and ſmooth ſtuffing than cotton or tow, which on that account are preferable. The reader will obſerve the difficulty and expence of following this complex method recommended by Kuckahn; it is indeed ſurpriſing that his prolix directions ſhould be admitted at large into the Philoſophical Tranſactions.

(*z*) It muſt be almoſt impracticable to proportion an artifical breaſt exactly of the natural ſize and ſhape; cotton or tow anſwers every purpoſe with leſs trouble;

by

by a longitudinal incifion from the breaft to the vent, not only the brain and flefhy parts, but likewife all the bones, except thofe of the legs, and the arch of the fkull, fhould be removed: the fkin is then to be ftuffed with a fufficient quantity of cotton or tow, to give the fubject its natural fize.

Were the infide furface of the fkin to be fprinkled with the dry antifeptic powder before-mentioned, or with tobacco-duft, it might probably be fome prefervative againft infects; but it is not at this time an ufual practice, experience having fhewn that the removal of the flefh and bones affords the beft fecurity.

To exhibit the bird in its proper attitude, there fhould be fome fubftitute for the back-bone, in order to give firmnefs to the fubject; for this purpofe brafs-wire nealed, (that is heated, in the fire, which makes it pliable and retain any form into which it is bent), or a thin piece of lead, may be paffed down the throat, or through one of the noftrils, quite to the vent; this fhould be done, before, or at the time of introducing the cotton or tow. Wires muft next be introduced through the feet and legs, or between the bone and fkin of the latter, and faftened to the longitudinal wire or lead, paffing from the head to the vent; the other ends of the wire may be fixed in the perch or board, on which the fubject is to remain.

Having then carefully fewed up the fkin, and placed the bird in the attitude we purpofe it to retain; it fhould next be expofed to a dry air, or

 to

to the warmth of an oven heated to a very moderate degree, and it will ſoon acquire a firmneſs which it will afterwards retain; and when this is effected, it ſhould be fixed in a box, ſecured from inſects, glazed in front to exhibit the preparation to view.

When perſons are on long voyages, they have ſeldom leiſure or convenience to go through this proceſs; and indeed were they to effect it, it would be difficult to preſerve the ſubject from inſects.

Under any embarraſſment of this kind, birds and other animals, particularly ſmall ones, may be put into brandy, wine, arrack or firſt runings, without being ſkinned; and the whole proceſs of preparation juſt deſcribed, may be executed when the traveller arrives in port.

Large Sea-fowl have thick ſtrong ſkins, and ſuch may be ſkinned; the tail, claws, head, and feet, are carefully to be preſerved, and the plumage ſtained as little as poſſible with blood.

Baking in a mild heat is not only uſeful in freſh preſervations; but, will alſo be of great

(*a*) Linné deſcribes another method of preſerving fiſh; this is to expoſe them to the air, and when they acquire ſuch a degree of putrefaction, that the ſkin loſes its coheſion to the body of the fiſh, it may be ſlid off almoſt like a glove; the two ſides of this ſkin may then be dried upon paper like a plant,. or one of the ſides may be filled with plaſter of Paris, to give the ſubject a due plumpneſs. Vid. Amæn. Acad. Vol. III. A fiſh may be prepared, after it has acquired this degree of putrefaction, by making a longitudinal

ſervice

ſervice to old ones, in deſtroying the eggs of inſects, if they be ſuſpected.

Small quadrupeds, all kinds of reptiles, as ſnakes, lizards, and frogs; fiſh (*a*) of all ſorts and ſmall tortoiſes, with ſea eggs (*b*), and ſea ſtars, may be put into brandy, rum, arrack, or firſt runnings, with the addition of a little alum.

Shells conſtitute an extenſive part of natural hiſtory, and may be collected in great plenty and variety on the ſhores of moſt iſlands and continents. Thoſe which are found with the fiſh in them, are the moſt valuable for the brightneſs of colour, and ſmoothneſs of ſurface, as they loſe that beauty and poliſh, when they have been long expoſed to the ſun. In bivalves, or thoſe having double ſhells, as cockles, oyſters, &c. both the ſhells ſhould always be collected. It is ſufficient in packing up ſhells, to prevent their rubbing againſt each other, which may be effected by means of paper, moſs, ſand, &c. Some of the ſhell-fiſh may be preſerved in ſpirits, as this might prove an uſeful addition to the knowledge of this department of natural hiſtory.

incifion on the belly, and carefully diſſecting the fleſhy parts from the ſkin, which are but ſlightly attached to it in conſequence of the putreſcency; the ſkin is then to be filled with cotton and tow, as directed for birds, and laſtly to be ſewed up where the inciſion was made.

(*b*) The echini, or ſea eggs, may alſo be dried; they are however liable to be broken.

The

The neſts and eggs of birds, would like-wiſe contribute to increaſe the knowledge of natural hiſtory, and prove alſo highly orna-mental among collections in that branch of zoology.

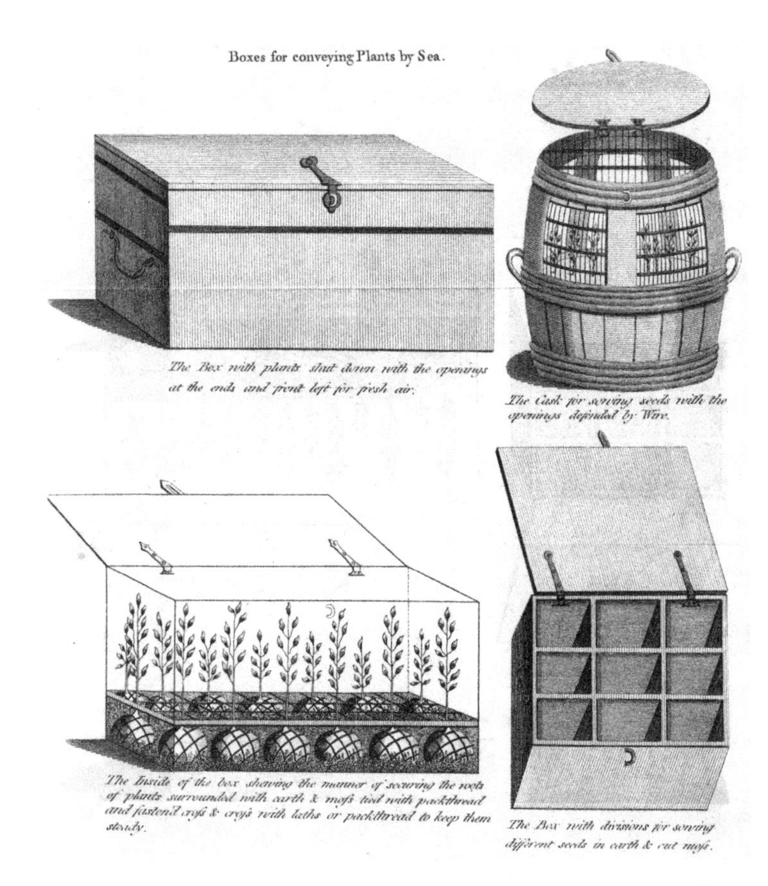

The material originally positioned here is too large for reproduction in this reissue. A PDF can be downloaded from the web address given on page iv of this book, by clicking on 'Resources Available'.

SECT. III.

Directions for bringing over Seeds *and* Plants *from distant Countries.*

Nor ev'ry plant on ev'ry soil will grow;
The sallow loves the watry ground, and low;
The marshes, alders: Nature seems t' ordain
The rocky cliff for the wild ash's reign;
The baleful yew to northern blasts assigns;
To shores the myrtles; and to mounts the vines (*c*)

EVERY part of the world has its peculiar productions; and in no objects of natural history, is the variety more entertaining, and important than horticulture.

The gardens of the curious have already been enriched with many valuable acquisitions from distant countries; but many attempts also to introduce several other plants equally rare, have been unsuccessfully made, owing to the bad state of the seeds or plants when first procured, or the method of disposing of them during long

(*c*) Nec vero terræ ferre omnes omnia possunt.
Fluminibus salices, crassisque paludibus alni
Nascuntur, steriles saxosis montibus orni.
Littora myrtetis lætissima: denique apertos
Bacchus amat colles; aquilonem et frigora taxi.
VIRG. G. II. 109.

voyages;

voyages; and ſuch accidents as the utmoſt precaution cannot prevent.

For the purpoſe of tranſportation, ripe ſeeds ſhould be choſen, which have been collected in dry weather, and kept dry without expoſing them to ſunſhine; and internally they ſhould be plump, white, and moiſt.

Attention to the ſtate of the ſeeds of Chineſe plants is peculiarly requiſite, as there is reaſon to ſuſpect that more ſeeds from China miſcarry from the art and treachery of the natives, than from the diſtance, or any defect in conveying them; as many ſeeds are brought over, which appear to have been roaſted by the Chineſe, previous to diſpoſing of them, in order to prevent their vegetation, and thereby keep up the demand. To diſcover the healthy ſtate of the ſeeds, ſome of the larger ones may be cut acroſs, and the ſmaller ones bruiſed, and by means of a magnifying glaſs, or even by the naked eye, it may be diſcovered, whether their internal part which contain the ſeminal leaves, appear plump, white and moiſt. If ſo, we may conclude they poſſeſs a vegetating ſtate; but if they are ſhrivelled, inclining to brown or black, and are rancid, they cannot in the leaſt be depended upon.

(a) Seeds thus carefully ſelected, may be preſerved by rolling each in a coat of yellow bees-wax, about half an inch thick; and afterwards a number of theſe, thus prepared, may be put into a chip-box, which is to be filled with melted bees-wax, not made too hot: the outſide of the box may then be waſhed with a ſolution of of ſublimate mercury (*d*), and kept during the paſſage in a cool airy place. In this manner tea-ſeeds, the ſtones of mangoes, and all hard nuts and leguminous ſeeds in general, may be prepared.

(b) Inſtead of putting ſmall ſeeds in bees-wax, they may be encloſed in paper or cotton which has been firſt ſteeped in melted bees-wax, and then placed in layers in a chip-box, ſome of which may be filled as before with melted bees-wax. Pulpy ſeeds, as thoſe of ſtrawberries, mulberries, arbutuſes, &c. may be ſqueezed together and dried, and then put into the cerate-paper or cotton above-mentioned.

(c) Small ſeeds, may alſo be mixed with a little dry ſand, put into the cerate-paper or cotton, and packed in glaſs-bottles, which are to be well corked, and covered with a bladder or leather (*e*).

(*d*) Sublimate mercury is the moſt effectually diſſolved in the acid of ſea-ſalt; one drop of which will diſſolve one grain of mercury, which will afterwards mix with water. One drachm of ſublimate will be ſufficient for half a pint of water.——Corroſive ſublimate may likewiſe be diſſolved in a ſaturated ſolution of ſal ammoniac in water, one ounce of which will diſſolve twenty ſcruples of ſublimate.

(*e*) This may be compared with what Dr. Hawkeſworth obſerves in his Voyages. vol. ii. p. 123. "On the 10th I put ſome ſeeds of melons, and other plants, into a ſpot of ground which had been turned up for the purpoſe: they had all been ſealed up, by the perſon of whom they were bought, in ſmall bottles, with roſin, but none of them came up except muſtard; even the cucumbers and melons failed, and Mr. Banks is of opinion that they were ſpoiled by the total excluſion of freſh air." Some ſeeds which I received from North America, encloſed in corked bottles, have ſince been ſown, and germinated. See g.

Theſe bottles may be put into a keg, box, or any other veſſel, filled with four parts of common ſalt, two of ſaltpetre, and one part of ſal ammoniac; or common ſalt alone, if the others cannot be procured, in order to keep the ſeeds cool, and preſerve their vegetative power.

(d) Seeds and nuts, in their pods, may be encloſed in linen or writing-paper, and put into caniſters, earthen-jars, ſnuff-boxes, or glaſs-bottles; the interſtices between the parcels of ſeeds ſhould be filled with whole rice, millet, panic, wheat-bran, or ground Indian-corn well dried. To prevent any injury from inſects, a little camphor, ſulphur, or tobacco, ſhould be put into the top of each canniſter or veſſel, and their covers well ſecured, to exclude the admiſſion of the external air.

(e) Seeds, well dried, may be put into a box, not made too cloſe, upon alternate layers of moſs, in ſuch a manner as to admit the ſeeds to vegetate, or ſhoot their ſmall tendrils into the moſs. In the voyage the box may be hung up at the roof of the cabin, and when the ſhip is at the place of her deſtination, the ſeeds ſhould be put into pots of mould, or boxes, with a little of the moſs alſo about them, on which they had lain.

(f) After every other precaution in tranſporting ſeeds has failed, I have known inſtances of their having been brought from diſtant parts, even from Botany-Bay, and Norfolk-Iſland, by the circuitous voyage of the Eaſt-Indies, in a perfect ſtate of vegetation, which have been merely wrapped in common brown-paper; and as it is the

the easiest and perhaps best method of conveying seeds, it should never be neglected: American seeds are usually brought over in this manner. The Chinese paper is generally employed for seeds from the East-Indies, and is probably as good as our common blotting and brown-paper.

Seeds preserved after the manner (e), (f), as well as that of (d), and likewise, for further security, some of the preceding (a), (b), (c), which have been procured in the East-Indies, may be examined when the ship arrives at St. Helena; and some of them, which appear in a state of vegetation, should be sown in the annexed boxes of earth, between the growing plants, as many sorts as possible; some of which may succeed in case of failure of the plants.

More of the same seeds may be also sown after the ship has passed the Tropic of Cancer, near the latitude of thirty degrees north. And if very small bits of broken glass are mixed with the earth, or thrown plentifully over its surface in the boxes, it may prevent mice and rats from burrowing in it, and destroying the tender roots of the plants and growing seeds.

In whatever method our seeds have been preserved, it should be a constant precaution to sow them as soon as they have been exposed to the external air, otherwise they probably will never vegetate.

In order to take up plants or shrubs advantageously, that are to be transported, a mattock and a spade should be provided; with the mattock a small trench should be opened round the plant intended to be taken up; the spade

 should

ſhould then be put under the root, which muſt be lifted up with a large ball of earth ſurrounding it; and if it ſhould fall off, it muſt be ſupplied with more earth, ſo as to form a ball about the roots of each plant, which muſt be ſurrounded with wet moſs, and carefully tied about with pack-thread to keep the earth about the roots moiſt: loamy earth will continue moiſt the longeſt.

Of each kind the youngeſt plants of ſhrubs and trees that can be found, ſhould be taken; none of them ſhould be above a foot high; as young plants are found by experience to bear removing much better than old ones.

Convenient boxes for the conveyance of plants in long voyages, are made about four feet long, two broad, and two deep; theſe, when half filled with earth, can be conveniently carried by two men holding the rope-handles fixed to their ends. pl. III. fig. 1. 2. 3.

Theſe ſhould be filled about half full of mould, with a few rotten ſticks or leaves at the bottom, and the plants intended to be ſent, planted in it, as ſoon after the ſhip's arrival as poſſible. When the ſhip is about to ſail, and they are ſent on board, hoops are to be nailed to the ſides of the box, in ſuch a manner, that, arching over it, they may cover the higheſt of the plants; ſmall ropes are to be twiſted between theſe, in the form of a net, to prevent the dogs or cats from getting at them, and ſcratching them up, on account of the freſh mould.

For each box ſo hooped and netted, provide a canvas cover, which may, when put on, entirely

protect

BOXES *for conveying* PLANTS *by Sea.*

Fig. 1.

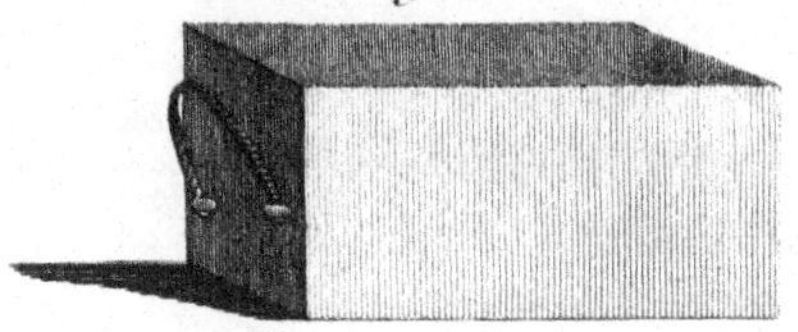

Fig. 2.

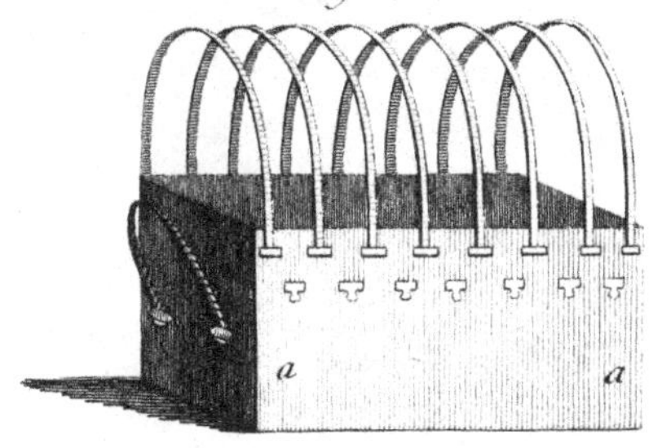

Fig. 3.

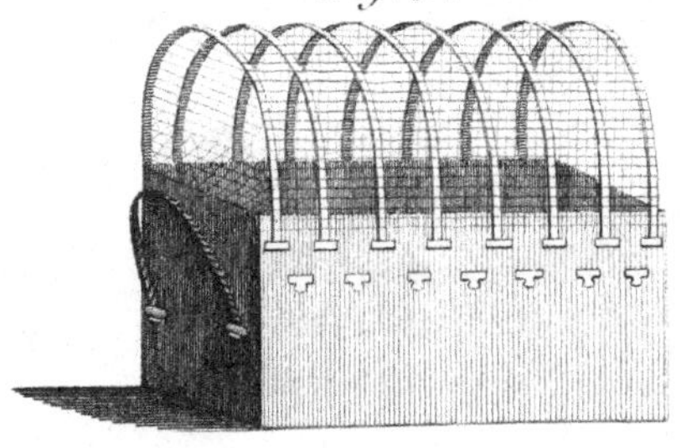

F. 1. *Form of the Box.*

2. *The same with hoops and loops. a. a. for securing the Canvas.*

3. *The same netted.*

J. Lodge sc.

protect it; and, to prevent this cover from being loſt or miſlaid, nai it to one ſide, and fix loops or hooks to the other, by which it may occaſionally be faſtened down.

The late J. Ellis, Eſq. who took ſingular pains to promote the introduction of Exoticks, recommends the conſtruction of a box and caſk agreeable to pl. IV. fig. 1. 2. The firſt ſhews the inſide of the box, and the manner of ſecuring the roots of plants ſurrounded with earth and moſs tied with pack-thread, and faſtened croſs and croſs with laths or packthread to keep them ſteady.

There muſt be a narrow ledge nailed all round the inſide of the box, within ſix inches of the bottom, to faſten laths or packthread, to form a kind of lattice-work, in order to ſecure the plants in their places, as abovementioned.

The caſk fig. 2. is convenient for ſowing of ſeeds, with the openings defended by wire; and as every ſhip has ſpare caſks, one may be readily formed for the conveyance of ſeeds in vegetation or growing plants, agreeable to the annexed engraving. The following is the proportion it ſhould be of; two feet three inches high, two feet bung diameter, and one foot nine inches head diameter; there ſhould be a large opening at the top wired over, the wired part of which might be lifted up at pleaſure, and a lid with hinges to cover it; this may be either circular or ſquare, as will be moſt convenient, the larger the better; and on the upper part of the ſides there may be four or five little openings wired, with doors to each, for the ſake of giving air all round upon ſome occaſions.

Care

Care muſt be taken not to expoſe the young plants to ſtrong ſunſhine: ſometimes, when the lid and doors are open, it may be neceſſary to throw a matt or thin cloth over them, but this muſt depend on the judgment of the perſon who has the care of them; there ſhould be handles fixed to the ſides, to move it with more ſafety.

There ſhould be a layer of wet moſs, of two or three inches deep, at the bottom of the box or caſk; or, if that cannot be got, ſome very rotten wood or decayed leaves, and then freſh loamy earth, about twelve inches deep, both of which will ſink to a foot deep: the wet moſs is intended to retain moiſture, and to keep the earth from drying too ſoon.

The ſurface of the earth ſhould be covered with moſs cut ſmall, which now and then on the voyage ſhould be waſhed in freſh water, and laid on the earth again to keep the ſurface moiſt, and to waſh off mouldineſs or ſaline vapours which may have ſettled on it. When the plants come up, it will be proper to ſave what rain-water can be got, which will encourage their growth, and be of more ſervice than the water drawn out of caſks that have been long on board the ſhip.

Theſe kind of boxes or caſks will be very proper to ſow many ſorts of ſuch ſeeds in, as are ſo difficult to be brought from China, and other parts of the Eaſt-Indies, to Europe in a vegetative ſtate; ſuch as the lechee, mangoes, mangoſteens, pepper, marking-nuts, various ſorts of peaches, roſes, oranges, citrons, lemons, &c.

If

If the plants ſent from theſe countries were planted in pots or boxes, and kept there a year, they might be brought over with very little hazard; or even if they were firſt tranſplanted from the woods into a garden, till they had formed roots, they might be ſent with much more ſafety.

The captain who takes charge of them, muſt be particularly informed, that the chief danger plants are liable to in ſea-voyages, is occaſioned by the minute particles of ſalt-water with which the air is charged, whenever the waves have white frothy curls upon them; theſe particles fall upon the plants, and quickly evaporating, leave the ſalt behind, which, choaking up the pores, prevents perſpiration, and effectually kills the plant; he therefore ſhould never let the covers be off, except on days when the wind is not ſufficiently high to beat the water up into what the ſeamen call white-caps. He muſt not keep them always ſhut up during the voyage; for if he does, they will mould and periſh by the ſtagnation of the air under the covers: and if at any time, by accident or neceſſity, they ſhould have been expoſed to the wind when the waves have white-caps, he muſt be deſired to water them well with freſh water, ſprinkling all the leaves with it, to waſh off the ſalt-drops which cover them. In this manner plants may be brought from almoſt any diſtance; many come from China every year in a flouriſhing ſtate.

If it be convenient to the captain to give up a ſmall part of the great cabin to the plants, this is certainly by far the beſt ſtation for them; nor

are

are they much in the way as the place which ſuits them beſt is cloſe to the ſtern windows; in this caſe they need not be furniſhed with their canvas covers; and they may frequently have air, by opening the windows when the weather is quite moderate.

As the Chineſe ingraft many plants which they introduce into their dwelling houſes, and which may be purchaſed in a healthy ſtate of vegetation, it is a neceſſary caution, not to earth ſuch plants in the paſſage too high with mould, as they are thereby liable to rot, or be otherwiſe injured by ſuch means. The rolling of the ſhip has ſometimes removed the earth, in which they are planted, and many fine collections have been deſtroyed by replacing the earth too high on the ſtems, particularly of ſuch plants as have been engrafted, a practice very general with the Chineſe.

From the frequent inconvenience of the rolling of ſhips, which ſometimes unearths the plants, or otherwiſe injures them, were the boxes or pots to ſtand upon a ſtage, moveable like the mariner's compaſs, it would certainly afford the beſt conveyance: this was prepared at conſiderable expence, in the veſſel commanded by Captain Bligh, to convey the bread-fruit tree, and other vegetables from the South-Seas: the plants he had procured were in high health, when this laudable and princely plan was fruſtrated by a mutiny on board, to be lamented by every man of ſcience and philanthropy. In merchants ſhips, where commerce is the object of the voyage, and where ſpace and conveni-

ence

ence can ſeldom if ever be commanded; and therefore the plan of a moveable ſtage cannot here be expected: from a munificent prince the combination of a company, or a ſociety, we muſt look for the ſupport of ſuch an enterprize.

One method of procuring plants from China has been ſucceſsfully adopted by that liberal promoter of horticulture Gilbert Slater Eſq. which may be introduced here. He has ordered every curious drawing of Aſiatic plants by Chineſe artiſts to be procured for him, in the power of ſea-faring gentlemen; theſe drawings are in general not only elegant and accurate; but ſeverally diſtinguiſhed by their names in Chineſe characters: the characters are copied on Chineſe paper, which being thin and almoſt tranſparent, may be accurately traced, and ſuch as refer to rare productions, are diſtributed to captains of ſhips, and other perſons undertaking voyages, with directions to purchaſe ſuch living plants, as correſpond with the character. Sometimes theſe are accompanied with the drawings themſelves. By this circuitous but advantageous mode, in a few years he will acquire a collection of valuable plants, at preſent not known in Europe.

It is much to be regretted, that after the moſt ſcrupulous attention in collecting and tranſporting very valuable plants, they ſhall ultimately be loſt by the tedious proceſs of our Cuſtom-houſe, employed to prevent ſmuggling. I once ſaw ſome fine plants, for which I ſhould have deemed fifty guineas a cheap purchaſe, totally ſpoiled by being kept there a full month, without any care whatever, not receiving the leaſt ſprinkling of water, whilſt the Thames flowed within twenty yards of their priſon. This does not reſult from any malicious deſign of the officers, who in their vigilance to prevent ſmuggling, thus occaſion the loſs of property, that might not only gratify laudable curioſity, but enrich our poſſeſſions with new products of importance in food, or arts. At this moment I am ſuffering under the injury that a collection of Aſiatic inſects have ſuſtained, of which I do not expect to receive one in a perfect ſtate.

A plan might be adopted, not only to obviate this painful inconvenience, but at the ſame time encreaſe rather than leſſen the revenue, by giving power to the commiſſioners of the cuſtoms to appoint gentlemen of taſte and knowledge in natural hiſtory, whoſe valuation of ſuch ſubjects ſhall govern the duties, which are uſually taken advalorem; and on payment, that the proprietors be permitted to remove their property at pleaſure. That gentlemen might be found to make theſe valuations gratuitouſly I doubt not; and as to the revenue on ſuch ſubjects, I know it would not ſuffer, for many things are, at preſent, not only ſpoiled or totally deſtroyed, but much undervalued by ignorant perſons, many proofs of which I have in my recollection.

When the naturaliſt is in ſearch of vegetable productions, different ſoils and ſituations ſhould be examined; as the ſea, and its ſhores, deep running waters, dikes, marſhes, moors, mountains,

tains, cultivated and barren fields, woods, rocks, &c. afford each their peculiar plants; and where-ever any are collected, the particular ſoil and ſituation ſhould be remarked. Sometimes it may prove inconvenient to convey the plants which may be diſcovered, when it would not be ſo to ſend them dried, in the form of a hortus-ſiccus. To do this in the beſt manner, and to make their ſtalks, leaves, &c. lie flat and ſmooth, " The plants ſhould be gathered in a dry day, " after the ſun hath exhaled the dew; taking " particular care to collect them in that ſtate " wherein the generic and ſpecific characters are " moſt conſpicuous; the ſpecimens ſhould be " ſuffered to lie on a table until they become lim-" ber, and then they ſhould be laid upon a paſte-" board, as much as poſſible in their natural " form, but at the ſame time with a particular " view to their generic and ſpecific characters: " for this purpoſe, it will be adviſeable to ſepa-" rate one of the flowers, and to diſplay the ge-" neric character: if the ſpecific character de-" pends upon the flower, or upon the root, a " particular diſplay of that will be likewiſe ne-" ceſſary. When the plant is thus diſpoſed upon " the paſteboard, cover it with eight or ten layers " of ſpongy paper, and put it into the preſs (*f*).

(*f*) " The preſs may be prepared by the following directions. " Take two planks of a wood not liable to warp, two inches " thick, eighteen inches long, and twelve inches broad. Get " four male and four female ſcrews, ſuch as are commonly uſed " for ſecuring ſaſh windows. Let the four female ſcrews be let

" Exert only a ſmall degree of preſſure for the
" firſt two or three days; then examine it, un-
" fold any unnatural plaits, rectify any miſtakes,
" and after putting freſh paper over it, ſcrew the
" preſs harder. In about three days more, ſe-
" parate the plant from the paſteboard, if it is
" ſufficiently firm to allow of a change of place;
" put it upon a freſh paſteboard, and covering it
" with freſh bloſſom-paper, let it remain in the
" preſs a few days longer. The preſs ſhould
" ſtand in the ſunſhine, or within the influence
" of a fire.

" When it is perfectly dry, the uſual method
" is to faſten it down with paſte or gum-water (*g*),
" on the right-hand inner page of a ſheet of

" into the four corners of one of the planks, and correſpond-
" ing holes made through the four corners of the other plank,
" for the male ſcrews to paſs through, ſo as to allow the two
" planks to be ſcrewed tightly together. It will not be amiſs
" to face the bearing of the male ſcrews upon the wood with
" iron plates; and if the iron plates went acroſs from corner
" to corner of the wood, it would be a good ſecurity againſt
" the warping."

This note I have copied from Dr. Withering's Botanical arrangement, and likewiſe the account of drying plants, as his directions are more full and complete than thoſe I formerly introduced into my Naturaliſt's Companion, An. 1772. Where the convenience of this preſs is not at hand, a ſuitable preſſure may be made by weights, or any heavy body.

(*g*) " A ſmall quantity of finely-powdered arſenic or corroſive
" ſublimate is uſually mixed with the paſte or gum-water, to
" prevent the devaſtations of inſects; but the ſeeds of ſtaves-
" acre, finely powdered, will anſwer the ſame purpoſe, without
" being liable to corrode, or to change the colour of the more
" delicate plants.'

" large

" large ſtrong writing-paper. It requires ſome
" dexterity to glue the plant neatly down, ſo that
" none of the gum or paſte may appear to defile
" the paper. When it is quite dry, write upon
" the left-hand inner page of the paper, the name
" of the plant, the ſpecific character, the place
" where, and the time when, it was found; and
" any other remarks that may be thought pro-
" per. Upon the back of the ſame page, near
" the fold of the paper, write the name of the
" plant, and it will then be complete for the
" cabinet."

" Some people put the dried plants into ſheets
" of writing paper, without faſtening them down
" at all; and others only faſten them by means of
" ſmall ſlips of paper, paſted acroſs the ſtem or
" branches.

" Another more expeditious method is, to take
" the plants out of the preſs, after the firſt or
" ſecond day; let them remain upon the paſte-
" board; cover them with five or ſix leaves of
" blotting-paper, and iron them with a hot
" ſmoothing iron, until they are perfectly dry:
" if the iron is too hot it will change the co-
" lours; but ſome people, taught by long practice,
" will ſucceed very happily. This is quite the
" beſt method to treat the orchis, and other ſlimy
" mucilaginous plants.

" Another method is, to take the plants when
" freſh gathered, and, inſtead of putting them
" into the preſs, immediately to faſten them down
" to the paper, with ſtrong gum-water; then dip

"a camel-hair pencil into ſpirit-varniſh (*h*), and "varniſh the whole ſurface of the plant two or "three times over. This method ſucceeds very "well with plants that are readily laid flat; "and it preſerves their colours better than any "other."

However beautiful a collection of dried plants may appear in the form of a hortus-ſiccus, yet where duplicates can be procured, it would be acceptable to receive plants both in flower, and in ſeed, dried in a careleſs manner, without nicety in expanding their foliage: by this means ſome of the flowers have been preſerved more entire; and afforded the botaniſt the moſt accurate characters of the plant, which by expoſure to the vapour of hot water, or being ſoaked in lukewarm water itſelf, has expanded, and exhibited the parts of fructification in the moſt perfect ſtate.

The ſeeds of a plant collected when they are ripe, will in a hortus-ſiccus, long retain their vegetative powers, and many of our valuable plants have thus been caſually introduced. It is well known, that the firſt green-tea-tree poſſeſſed by this country, was raiſed by the late John Ellis Eſq. from a ſeed picked out of a canniſter of tea.

(*h*) The ſpirit-varniſh may be made of a quart of highly-rectified ſpirit of wine, five ounces of gum ſandarach, two ounces of maſtich in drops, one ounce of pale gum elemi, and one ounce of oil of ſpike lavender: theſe are to ſtand in a warm place, and be ſhook frequently, to expedite the ſolution of the gum.

The

The impreſſions of plants well taken off upon paper, look very little inferior to the beſt drawings, and may be done with very little trouble. For this purpoſe, ſome printer's ink (*i*), and a pair of printer's balls, ſuch as are uſed for laying the ink on types, are neceſſary. After rubbing theſe balls with a little of the ink, lay the plant betwixt them, and preſs it ſo as to give it ſufficient colour; then take the plant and lay it carefully on a ſheet of paper, and preſs it with the hand, to give the impreſſion of the plant to the paper, which may be afterwards coloured according to nature; a piece of blotting-paper may be placed betwixt the plant and the hand, to prevent the latter from being dirtied by the ink.

An effectual method of ſending a branch of any plant, with the flowers and parts of fructification entire and perfect, is to put them in bottles of brandy, rum, or arrack; but the colour of the plant is often injured.

Corals, corallines, ſponges, &c. inhabitants of the ſea, are found in conſiderable variety near the coaſts of iſlands and continents, particularly in hot climates. Some of theſe are very tender and brittle when dry, and ſhould therefore be carefully packed up in ſand, in order to keep them ſteady, or placed betwixt papers in the manner of an hortus-ſiccus.

In hot climates, the inſects are very rapacious and I have ſeen the fineſt fan-corals, and others of a ſoft texture when firſt taken out of the ſea,

(*i*) Where this cannot be procured, ivory, or lamp-black, ground with boiled linſeed-oil, may be ſubſtituted.

almoſt

almoſt devoured by ants, before they became dry and hard. To prevent injuries of this kind, a little powdered corroſive ſublimate, arſenic, or ſtaveſacre, may be ſprinkled upon theſe productions. Some of the ſmall, and branches of the large ones, might be alſo put into ſpirits, and the parts of them thereby preſerved much more diſtinct; which would ſerve greatly to illuſtrate their natural hiſtory.

SECT. IV.

Method of Analyzing Mineral Waters (a).

Qui autem ad obſervandum adjicit animum, ei etiam, in rebus quæ vulgares videntur, multa obſervatu digna occurrunt.

Bacon, de Augment. Scient.

AS many ſprings contain a volatile principle, ſoon liable to be diſſipated, it is neceſſary to make our experiments on the ſpot, in order to diſcover the contents of ſuch waters. Various as theſe contents may appear, the apparatus proper to detect them, may be reduced into a ſmall compaſs.

When we purpoſe to examine any mineral or medicinal ſpring, the ſoil and face of the country ſhould be conſidered, the ſtony or mineral appearance, and particularly whether there are any mineral veins: the degrees of heat of the water ſhould be aſcertained by a thermometer, and its comparative weight to other ſprings in

(*a*) Wallerius in his Hydrologia, Lewis in his notes on Neumann's Chemiſtry, with Rutty, Lucas, Falconer, Monro, Pearſon, Garnett, and other Engliſh writers, have given directions on this ſubject, which ſome late chemical writers on the continent have further elucidated, as Wiegleb, Bergman, Strachling, Struve, Weſtrumb, and Gottling.

the neighbourhood alſo carefully obſerved; after which we may enter on our experiments.

From the ſubſtances contained in mineral waters, they admit of the four following diviſions.

I. ACID AND NEUTRAL SALTS.
II. EARTHLY SALTS.
III. METALLIC SALTS.
IV. SULPHUREOUS.

I. ACID AND NEUTRAL SALTS.

More eminently ſaline, containing the acids, and the compound of acids and alkalis forming neutral ſalts.

1. Fixed air * the only acid found in any quantity pure.
2. Aerated ſoda.
3. Glauber's ſalt.
4. Vitriolated tartar.
5. Saltpetre.
6. Marine ſalt.

†

II. EARTHY SALTS.

1. Epſom ſalt.
2. Magneſian nitre.

* Vitriolic acid is found in extremely ſmall quantity, and that very ſeldom.

† Perhaps Borax, but it is not quite certain that this is a Mineral.

3. Mag-

3. Magnesian muriate.
4. Magnesia, or common Magnesia.
5. Gypsum.
6. Calcarious nitre.
7. Calcarious muriate.
8. Chalk.
9. Alum.

III. METALLIC SALTS.

1. Aerated iron.
2. Vitriolated iron.
3. Vitriolated copper.
4. Vitriolated zinc.
5. Arsenic.

IV. SULPHUREOUS.

1. Hepatic air.
2. Liver of sulphur.

Water is found mixed with some few substances not here enumerated, as

a. Silicious earth in very fine particles sometimes gives to water a milkiness, and

b. Argillaceous earth in such quantity as to give the water the title of saponaceous, from its feel and resemblance.

Some other substances, as bitumen and naptha, may give Water sometimes a little character, as they flow out of the same springs upon the surface of some waters.

In this view of the substances contained in mineral waters, we see they are almost all saline,

 and

and as ſome of them are very generally diſtribut-ed through the earth, we find, unfortunately for our inveſtigation of this ſubject, that mineral water generally contains ſeveral of them, ſome-times many, which makes it neceſſary to ſubject portions of the ſame water to a variety of expe-riments.

Gypſum, calcarious muriate, chalk, Epſom ſalt, muriate of magneſia, and common magneſia, aerated iron, ſea-ſalt, and fixed air are among the moſt common ſubſtances found in mineral waters.

CLASS

CLASS I.

ACID AND NEUTRAL SALTS.

	Synonymous Names.
1. FIXED AIR.	Carbonic acid. Carbonaceous acid. Cretaceous acid. Mephitic acid. Aerial acid. Atmoſpheric acid. Spiritus ſylveſtris. Gas ſylveſtre, &c. &c. Gas.

This ſubſtance is found in every part of nature that we are acquainted with, in its different ſtates; fixed as it has been called, and aeriform. In our atmoſphere it is found in the proportion of about one in 100. Pits and cavities are frequently filled with it. It exiſts in immenſe quantities combined with calcareous earth, forming chalk and marble; and with many other ſubſtances. It is conſtantly thrown out from animals and vegetables; it is the great product of combuſtion, and formed in ſhort by every proceſs, in which pure air is combined with carbonaceous matter; hence it is no wonder that it is ſo general and ſo frequently found the uniting medium to ſubſtances contained in mineral waters. Mineral waters, which contain this alone, or conſiderably prevailing, have been called Acidulæ; and perhaps hence all mineral waters were formerly termed Acidulæ.

a. Fixed

a. Fixed Air is difcovered in water, by its fparkling, on agitation, and bubbles of air being given out from it.

b. By its pungent and acid tafte.

c. Tincture of litmus added to it, is changed from its blue colour to a red.

d. On its being added to lime water, white clouds are formed. The fixed air combining with the lime forms chalk, which not being foluble, is precipitated in a fine white powder.

e. By boiling it, the fixed air is expelled, and the quantity contained in any portion of water may be meafured, by conveying it through a tube into lime water, and collecting the precipitated chalk; every 30 grains of which contains 18 grains and ½ of fixed air.

2. AERATED SODA.

Mineral Alkali combined with fixed air.

	Synonymous Names.
2. AERATED SODA.	Marine alkali.
	Mineral alkali.
	Foffil alkali.
	Natron or natrum.
	Soda.
	Bafe of marine falt.
	Effervefent foda.
	Mephite of foda.
	Chalk of foda.
	Cretaceous foda.

This

This ſalt is found in large quantities in Aſia and Africa, and in their mineral waters, and in the waters of ſome parts of Germany; its cryſtals are commonly rhomboidal, extremely ſoluble in water, but when expoſed to the air, fall into powder, it is detected in water by

a. Paper ſtained with the tincture of litmus is changed to a blue colour.

Tincture of litmus is prepared by infuſing litmus, powdered, for a ſhort time, in diſtilled water; with this the paper is ſtained.

b. Paper ſtained with the tincture of brazil wood looſes its red, and is changed to a violet.

The tincture of Brazil wood is made in the ſame manner as the tincture of litmus, but hot water ſhould here be employed.

c. Paper, ſtained yellow by an infuſion of turmeric, is changed to a brown colour.

The infuſion of turmeric ſhould be made with hot water.

d. Upon adding a ſolution of corroſive ſublimate mercury to this water, a white precipitate is produced; if the alkali be in large proportion, this precipitate will be an orange colour. In this inſtance the alkali unites with muriatic acid of the corroſive ſublimate, and the mercury is precipitated.

e. Add a ſmall quantity of Epſom ſalt to this water, it will diſſolve, and no precipitation will be perceived, but give it a very little warmth, and immediately white flakes will be ſeperated: here the mineral alkali combines with the vitriolic acid of the Epſom ſalt, and detaches the magneſia its other conſtituent part, which alſo unites

to

to the fixed air of the mineral alkali. The mineral alkali gives out fixed air enough to rediſſolve it in the water; but expoſing it to a little heat detaches this fixed air, and the magneſia will be precipitated. This magneſia may be further diſſolved again with effervefcence, by adding any acid.

f. Upon adding a ſmall quantity of ſal. ammoniac, the mineral alkali will combine with the marine acid of the ſal ammoniac, and detach the volatile alkali, which will fly off, and may be rendered viſible by holding a paper moiſtened with ſome volatile acid over it, as the nitrous, marine, or acetous. Theſe forming nitrous, marine, or acetous ammoniac, which becomes more viſible.

g. By adding a ſmall quantity of a ſolution of blue vitriol in water, a precipitate of a bright apple green colour will be occaſioned by the alkali combining with the vitriolic acid of the blue vitriol, forming Glauber ſalt, which is ſoluble in the water, while the fixed air combines with the copper, which being inſoluble, is precipitated.

3. GLAUBER SALTS.

Natron combined with Vitriolic Acid.

	Synonymous Names.
3. GLAUBER SALTS.	Sal. mirabile.
	—- catharticus.
	Soda vitriolata.
	Natron vitriolatum.
	Sulphate of ſoda.

Its cryſtals are priſms with ſixth unequal and ſtriated ſides, terminating in dihædrel ſummits: it has a bitter taſte. When expoſed to the air, the cryſtals

cryſtals fall into powder, it is very ſoluble in water. This is found in abundance in ſea-water, and many mineral waters. It is detected by

a. A Solution of muriated ponderous * earth being added to water containing this ſalt, a precipitate, in the form of white clouds will immediately appear, and white precipitate will be gradually depoſited. The ponderous earth leaving the marine acid, will unite with the vitriolic, and form ponderous ſpar, which being inſoluble in water, gives this appearance when precipitated from it, of a white ſoft cloud, peculiar to itſelf: The marine acid combining with the natron, forms marine ſalt, which remains diſſolved in the water.

b. A ſolution of nitre of ſilver † being added, the water firſt becomes of an opal or milky hue, and after ſome time acquires a rediſh caſt, depoſiting at length a ſediment of that colour. Here the vitriolic acid combines with the ſilver, forming ſulphate of ſilver, a ſubſtance almoſt inſoluble in water, therefore precipitated; The nitrous acid, combining with the natron, forms cubic nitre, which remains diſſolved in the water.

c. A ſolution of lead combined with the acetous acid being added, a white percipitate will immediately be formed. The lead combines with the vitriolic acid, forming a ſubſtance inſoluble in water, while the acetous

* Ponderous earth combined with marine acid.

† Silver combined with nitrous acid.

acid, combining with the natron, remains diſſolved in the water.

d. An equal quantity of highly rectified ſpirit of wine being added to water containing this ſalt, the mixture becomes turbid, and by degrees ſmall cryſtals will be formed at the bottom of the veſſel; the ſpirits of wine combining with the water, ſeparates it from the vitriolated tartar, which is precipitated from it in the form of ſmall cryſtals.

4. VITRIOLATED TARTAR.

Kali combined with vitriolic Acid.

	Synonymous Names.
4. VITRIOLATED TARTAR.	Arcanum duplicatum. Sal de duobus. Sal polychreſtum. Kali vitriolatum. Nitrum vitriolatum. Sulphate of potaſh.

It has a bitteriſh taſte, ſuffers no alteration from expoſure to the air; the form of its cryſtals varies much, according to circumſtances under which they are prepared; they are moſtly ſix ſided pyramids, ſometimes ſix ſided priſms, like rock cryſtals: they are ſmall compared with the foregoing.

This ſalt is rarely found in mineral waters.

a. By adding a ſolution of muriated ponderous earth to water containing this ſalt, the ſame appearances take place as deſcribed (49 a.) while the

the muriatic acid combining with the vegetable alkali, forms sal digestivum sylvii, which remains dissolved.

b. By the addition of the mercurial solution (*c*) yellowish clouds will immediately be formed, and in time a precipitate of the same colour will be deposited. The mercury, leaving the nitrous acid, unites with the vitriolic, forming turpith mineral, which is insoluble in water, though a part of the vitriolic acid and mercury remain dissolved along with nitre, formed by the nitrous acid uniting with the vegetable alkali.

c. The addition of the solution of sugar of lead in water separates a white precipitate, as mentioned page 49. c. The vitriolic acid unites with the lead, while the acetous acid combines with the vegetable alkali, forming diuretic salt, which continues dissolved.

d. By adding an equal quantity of highly rectified spirits of wine, the mixture will become turbid, and by degrees small chrystals of vitriolated tartar will be formed at the bottom, for reasons assigned page 50. d.

(*c*) Mercury combined with nitrous acid by the assistance of heat.

SALT

5. SALT PETRE.

The compound of nitrous Acid with Vegetable Alkali.

Synonymous Names.

5. SALT PETRE. { Nitre. Common nitre. Prismatic nitre. Nitre of potash.

This is found generally near places inhabited by animals; it is inflammable, its crystals are six-sided prisms, terminating in dihedral pyramids, or cut with a slope, and frequently channelled.

a. When found in water, it may be detected, by heating the water containing it, then adding a little vitriolic acid to it, this unites with the alkali, forming vitriolated tartar, and detaches the nitrous acid, which being volatile, may be rendered sensible, by holding any thing moistened with volatile alkali over the surface; the vapour of the volatile alkali, uniting with the vapour of the nitrous acid, forms a visible whitish smoke, consisting of the nitrous ammoniac in fine dust.

b. By adding an equal quantity of highly rectified spirits of wine to this water, it becomes turbid, and this salt is separated from it, with the appearances as described page 50 d.

c. A piece of paper moistened with it, and suffered to dry, will be found to become touch-paper.

6. COMMON SALT.

A Compound of Muriatic Acid with Mineral Alkali.

Synonymous Names.

6. COMMON SALT.
- Salt.
- Sal gemmæ.
- Sal marinum.
- Gem ſalt.
- Sea or marine ſalt.
- Culinary or kitchen ſalt.
- Muriate of ſoda.

This ſalt abounds in nature, in vaſt maſſes in the earth; diſſolved in the water of the ſea; and in ſalt lakes; and, therefore, very frequently met with. Its cryſtals are regular tubes; it is detected

a. By adding vitriolic acid to the water containing it, this unites with the mineral alkali of the ſalt, and forms Glauber's ſalt, which remains diſſolved in the water; the muriatic acid is dětached, and, being a volatile ſubſtance, may be rendered viſible as the nitrous acid page 52. a. by means of the vapour of volatile alkali held over it; the volatile alkali uniting with the muriatic acid, forms ſal ammoniac, which becomes viſible in this very fine duſt.

b. By adding a ſolution of nitre of ſilver, a white precipitate is immediately formed, which ſome time after will acquire a bluiſh appearance, and this precipitate will not be rediſſolved by the addition

addition of either the nitrous or acetous acids; the ſilver uniting with the muriate acid of the ſalt, forms luna cornea, a ſubſtance ſcarcely at all ſoluble in water; the nitrous acid unites with the mineral alkali of the ſalt, forming cubic nitre, which remains diſſolved in the water.

c. Upon adding the mercurial ſolution, white clouds will immediately appear, and a precipitate of the ſame colour will ſubſide to the bottom.

In this inſtance the muriatic acid of the ſalt unites with the mercury, forming calomel and corroſive ſublimate; the calomel being almoſt inſoluble in water, is precipitated. The corroſive ſublimate remains diſſolved in the water, with the cubic nitre, formed by the nitrous acid and the mineral alkali.

CLASS II.

EARTHY SALTS.

1. Epsom Salt.

Vitriolic Acid combined with Magnesia.

Synonymous Names.

1. EPSOM SALT. { Bitter purging salt.
Magnesian vitriol.
Salt of sedlitz.
Sulphate of magnesia.

It was formerly obtained from wells in the neighbourhood of Epsom, and in the hills that run to the eastward of that place, whose waters contain it abundantly. It has a very bitter taste, is very soluble in water, and is very apt to form short needle-like crystals, when hastily crystalized; otherwise they are quadrangular prisms, terminating in quadrangular pyramids.

A. The magnesian basis may be detected

a. By the addition of lime water, to water containing this salt, which renders it turbid, and a flaky precipitate will be gradually formed. Lime has a stronger affinity to vitriolic acid than magnesia, combines with it and forms gypsum, a small portion of which remains dissolved in the water, the rest is precipitated along with the magnesia.

b. A

b. A ſolution of cauſtic vegetable alkali alſo renders it immediately turbid, and a white flaky precipitate is depoſited. The cauſtic vegetable alkali uniting with the vitriolic acid, forms vitriolated tartar, which remains diſſolved in the water, while magneſia being inſoluble, is precipitated.

c. But if a ſolution of mild vegetable alkali, or vegtable alkali combined with fixed air, be made uſe of, there will be no precipitation; or, if the liquor ſhould grow turbid, it will become clear again on being ſtirred the magneſia being rediſſolved by means of the fixed air that was contained in it. If the liquor be now placed in a warm ſituation, and this fixed air be ſuffered to eſcape, the magneſia will appear in a white precipitate.

d. By addition of the mineral or volatile alkalis, in their cauſtic and mild ſtate, ſimilar effects will be produced, as here mentioned, where vegetable alkali is uſed in its cauſtic and mild taſte.

e. A ſolution of ſoap being added to water containing Epſom ſalt, it becomes turbid, and a number of white flakes are depoſited. The alkali of the ſoap combines with the vitriolic acid, and forms vitriolated tartar, which remains diſſolved in the water, and the unguinous part with the magneſia forming an earthy ſoap.

f. A ſolution of acid of ſugar being added, no precipitation will take place, unleſs ſome calcarious earthy ſalt be mixed with it.

B. The nature of the acid contained may be detected

g. By

g. By adding a ſolution of ponderous earth in marine acid to this water, white clouds will be inſtantly formed, and a white precipitate be gradually depoſited (deſcribed page 49. a.) The magneſia combines with the marine acid, forming muriate of magneſia, which remains diſſolved in the water.

h. Or by adding a ſolution of nitre of ſilver, there will be no precipitate, but the liquor will have an opal appearance, and after ſometime a reddiſh one, depoſiting at laſt a precipitate of this colour. The magneſia, combining with the marine acid, remains diſſolved. The ſilver, combining with the vitriolic acid, forms this precipitate as deſcribed (page 49. b.)

i. The ſolution of nitre of mercury produces yellowiſh clouds, and by degrees a precipitate of the ſame colour, which is turbeth mineral (page 51. b.) The magneſia, uniting with the nitrous acid forms nitrate of magneſia, which remains diſſolved in the water.

k. If a ſolution of ſugar of lead be employed, a white precipitate, vitriol of lead, will be immediately formed by the lead uniting with the vitriolic acid. The magneſia unites with the acetous acid, and remains diſſolved.

100 Grains of vitriol of lead contains 28 grains of vitriolic acid.

l. Highly rectified ſpirits of wine produces a turbid mixture, and by degrees ſmall cryſtals of Epſom ſalt will be formed at the bottom of the veſſel.

m. A ſolution of blue vitriol being employed, produces clouds of a pale green colour, and after ſome time a precipitate of the ſame colour.

2. MAGNESIAN NITRE.

Nitrous Acid combined with Magneſia.

Synonymous Names.

2. MAGNESIAN NITRE. { Nitre of magneſia. Nitrated magneſia. Nitrate of magneſia.

It is found in the mother water of nitre, and, like it, very rarely found in mineral waters; it has an acrid and very bitter taſte; it imbibes moiſture from the air.

A. a. The magneſian baſis is detected by the ſame means, and with the ſame appearances, as (page 55, a.)

b. By lime water, the nitrous acid uniting with the lime, the magneſia is precipitated in a white flaky appearance. The cauſtic, vegetable, foſſile, or volatile alkali being added, the ſame appearances are produced, the water becomes turbid, and white flakes appear, as deſcribed (page 56, b.) The cauſtic alkali, uniting with the nitrous acid, precipitates the magneſia.

c. So alſo, when employed in their mild ſtate, the ſame appearance will take place, as deſcribed, (page 56, A. c.)

d. With the addition of ſoap, white flakes appear, as deſcribed above, the nitrous acid combining with the alkali of the ſoap, and the magneſia with the oil.

B. e. The nature of the acid may be detected, by adding a little vitriolic acid, when the nitrous acid will be detached, as described (page 52, a. and 53, a)

f. A solution of blue vitriol gives the same appearances as described (page 57, m.)

3.. MAGNESIAN MURIATE.

Muriate Acid combined with Magnesia.

Magnesian muriate, or muriate of magnesia, exists in sea water and salt spring waters, but in a more pure state in Epsom and similar waters, is therefore very common. Magnesian muriate has a very bitter hot taste, is very soluble in water, difficult to chrystalize ; its chrystals are needle shaped.

A. a. The magnesian basis may be detected in waters containing this salt, by the same means, and with the same appearances, as described (page 58, a. b. c. d.) The lime and alkalis, combining with the muriatic acid, detach the magnesia.

b. By means of a solution of soap in water, the same appearances take place, as described (page 58, d.)

B. c. The acid contained in this water may be detected by the means described (page 59, B. e.) The muriatic acid being volatile, is detached in an invisible vapour which uniting to the vapour of the volatile alkali, forms a whitish smoke, consisting of sal ammoniac.

d. By adding a little of the solution of silver in nitrous acid, white clouds will appear, ac-

 quiring

quiring in time a bluiſh appearance, as deſcribed (page 53, b.)

e. Or by adding a little of the ſolution of quickſilver in nitrous acid, white clouds will immediately appear, and the ſame appearances take place as deſcribed (page 53, c.)

f. Upon adding a little of the ſolution of blue vitriol, the precipitate of the pale green colour, above deſcribed, will appear.

4. MAGNESIA;

OR

Magneſia combined with fixed air.

Synonymous Names.

4. MAGNESIA.
- Magneſia alba.
- Common magneſia.
- Mild or efferveſcing magneſia.
- Magneſian earth.
- Aerated magneſia.
- Magneſian chalk.
- Muriatic earth.
- Cretacious magneſia.
- Mephite of magneſia.
- Carbonated, or carbonate of magnéſia.
- Magneſian carbonate, &c.

Magneſia is found no where pure, but, combined with other earths, it is found in abundance in nature, in a variety of forms: it is found in the mother water of nitre, and in the other ſtates mentioned in the two preceding pages.

A. a. The magnefia, contained in the water, may be detected by means of a folution of cauftic, vegetable, or foffile alkali. The cauftic alkali unites with the fixed air, is rendered mild, and remains diffolved in the water, while the magnefia is precipitated

b. By fufpending a piece of reddened* litmus paper in it, it will loofe the red colour given it by the vinegar, being blue again. And that this is only owing to the magnefia will be farther proved, by evaporating the water half away, the remaining liquor will be found to poffefs no power of changing the colour of the reddened litmus paper, and the fixed air having efcaped, the magnefia is precipitated in a white powder, and the liquor will be found to be water only.

c. If the piece of paper, that is made ufe of, be ftained with the tincture of Brazil wood, inftead of litmus, it will be changed from a red to a violet colour.

d. Or if a folution of foap be added, as mentioned (page 56, e.) The foap is decompofed, the alkali of the foap uniting with the fixed air, and the oil with the magnefia, forming an earthy foap.

B. e. The acid bafis will be detected to be fixed air by evaporation, in a clofe veffel, as defcribed (page 46, e.) The fixed air, as it flies off, may be forced into lime water, which it will decompofe.

* Firft ftain the paper with a little infufion of litmus, then wet it with vinegar.

GYPSUM

5. GYPSUM.

Vitriolic Acid combined with calcarious Earth.

Synonymous Names.

5. GYPSUM. { Vitriol of lime.
Calcarious vitriol.
Selenite.
Paris plaister.
Sulphate of lime.

This salt abounds in nature, and therefore is exceeding commonly found in spring water; but so small a quantity dissolves, that it gives very little taste to water.

A. a. In waters, containing Gypsum, the earthy basis may be detected by a piece of paper, stained with decoction of Brazil wood, being dipped into this water, will tinge it of a violet colour.

b. If a little acid of sugar be added to water containing it, the water will immediately become turbid, and a white powder be deposited. The acid of sugar has a stronger attraction to the calcarious earth than the vitriolic acid combined with it, and forms saccharine selenite, a salt extremely difficult of solution, therefore precipitated while the vitriolic acid remains diffused through the water.

c. A solution of the caustic vegetable alkali being added, produces white clouds. The calcarious earth is precipitated, in its caustic state, by the union of the alkali with its vitriolic acid, on on the water being ſtirred. This ſoon diſſolves again; cauſtic calcarious earth or lime being ſoluble in water, but by expoſing it to the air, or adding a little fixed air to it, it becomes chalk, and is again precipitated.

d. Or the mild vegetable alkali being employed, the precipitate will not diſſolve in the water, but fall to the bottom, and will be found to poſſeſs all the properties of common areated calcarious earth, or chalk.

e. So, inſtead of the vegetable alkali, the foſſile or volatile be employed in its cauſtic ſtate, no precipitation is produced, lime being precipitated; but if the mild volatile alkali be made uſe of, a white powder will be precipitated, which will be found to be chalk.

f. A ſolution of ſoap being added to this water, the ſame appearances will take place, as when added to water containing Epſom ſalt. The calcarious earth uniting with the oil, and being precipitated, the vitriolic acid uniting with the vegetable alkali.

B. g. The acid element is diſcovered by the ſolution of muriated ponderous earth; upon adding this, white clouds are immediately produced, as deſcribed (page 49) and a white inſoluble precipitate, which is an artificial ponderous ſpar, will be gradually depoſited at the bottom of the veſſel, as deſcribed (page 49); only here the calcarious earth remains diſſolved in water in union with the marine acid.

h. Upon adding the ſolution of nitre of ſilver, nitre of mercury, ſugar of lead, or blue vitriol, to waters containing Gypſum, the ſame appearances take place as enumerated (page 57).

6. CALCARIOUS NITRE.

Nitrous Acid combined with calcarious Earth.

Synonymous Names.

6. CALCARIOUS NITRE. { Calcarious nitrate. Mother water of nitre. Nitrate of lime.

It is very rarely found in waters; it, has a bitter disagreeable taste, with something of the taste of nitre.

a. By adding the vitriolic acid to water containing this salt, and warming the liquor, a vapour will arise, which may be rendered visible, by placing a stopper, moistened with volatile alkali, over the surface of this liquid. Here the vitriolic acid, combining with the calcarious earth, forms selenite, which, if in any considerable quantity, will be precipitated and form chrystals The nitrous acid is detached, and, being volatile, will fly off, and at that time becomes visible by the vapour of the volatile alkali, as described (page 52).

b. By the addition of the saccharine acid, the water will immediately become turbid, and a white powder will be deposited at the bottom of the vessel. The saccharine acid, in the same manner, unites with the calcarious earth, and forms calcarious selenite, which, being almost insoluble, is precipitated, while the nitrous is detached, and flies off.

c. A ſolution of cauſtic, vegetable, foſſil, or volatile alkali, unites with the nitrous acid, forming nitre, which remains diſſolved in the water, and precipitates the calcareous earth in the form of lime, which diſſolves again in the water, and may be detected as related (page 63).

d. And a ſolution of mild alkali, in like manner, unites with the nitrous acid, calcareous earth being precipitated in its mild form, or chalk, as mentioned in the ſame page.

e. On the addition of a ſolution of ſoap, the ſame appearances will take place, as when it is added to water containing Epſom ſalt (page 56. e.)

f. A ſolution of blue vitriol added to water containing this ſalt, clouds of a pale green colour will be formed.

CALCAREOUS MURIATE.

Muriatic Acid combined with calcareous Earth.

	Synonymous Names.
CALCAREOUS MURIATE.	Mother water of marine ſalt.
	Calcareous marine ſalt.
	Fixed ſal ammoniac.
	Muriate of lime.
	Murias calcareous.

It is found wherever ſea ſalt is met with, particularly in ſea water; chalk often contains it; it has a ſalt and very diſagreeable bitter taſte. It cryſ-

 talizes

talizes in prifms, with four ftriated faces, terminated with fharp-pointed pyramids. It deliquefces immediately on expofure to the air.

A. To water containing this falt, to detect the earthy bafis,

a. Add a little folution of the acid of fugar, the mixture becomes immediately turbid; the acid of fugar, uniting with the calcareous earth, forms a compound fcarcely foluble in water, which is precipitated, leaving the muriatic acid in the water, as defcribed (page 62, b.)

b. Add a little of the folution of either of the alkalis, in a cauftic ftate, white clouds will immediately form, the alkali uniting with the muriatic acid, the calcareous earth is precipitated; but here, on the liquor being ftirred, the precipitate difappears, for the calcareous earth being in its cauftic ftate (Lime) on being mixed with the water, diffolves in it; but if this liquor be expofed to the air, or if fixed air be thrown into it, this lime becomes chalk, which will fall to the bottom.

c. If the folutions of the alkalis be employed in their mild ftate, or combined with fixed air, white clouds will immediately form, as above, which will difappear again, on ftirring the liquor, but on another principle, there being a double affinity; for while the alkali combines with the muriatic acid, the fixed air, which was combined with the alkali, unites with the calcareous earth, and forms chalk; but here the fixed air is detached in fuch abundance, as alfo to diffolve the chalk fo formed in the water; now warm the water, the fuperabundant fixed

fixed air will fly off, and the chalk be precipitated.

d. By adding a ſolution of ſoap, the ſame appearances will take place, as in water containing Epſom ſalt, deſcribed (page 56, e.)

By adding a ſolution of ſilver in nitrous acid, or of mercury in nitrous acid, or of lead in acetous acid, or of blue vitriol, or by adding ſpirit of wine, the ſame appearances will be produced as upon adding theſe ſubſtances to water containing Epſom ſalts, deſcribed (page 57, h. i. k. l. m.)

B. e. The acid may be detected by the ſame means, and with the ſame appearances as deſcribed (page 53 and 54, a. b. c.)

CHALK.

Calcareous Earth combined with fixed Air.

Synonymous Names.

CHALK.	Creta. Cretaceous earth, &c. or Calcareous earth. Calcareous ſpar. Calcareous carbonate. Carbonate of lime. Aërated lime.

This is found in immenſe quantities in moſt parts of the earth; at leaſt the ſuperficial parts; of courſe is found diſſolved in almoſt all ſpring waters.

A. a. Upon placing a bit of reddened litmus paper in water containing it, the red colour will gradually difappear and the blue return; but if a portion of the water be firft evaporated, it will become turbid, the fuperabundant fixed air, which keeps the chalk in folution, will fly off, and the chalk be depofited: now place the litmus paper in it, and no alteration of colour will take place. The calcareous earth not being in fufficient quantity in the water to combine with the vinegar, that gives this red colour to the litmus paper.

b. Upon putting into it a piece of Brazil wood paper, the red colour will be by degrees changed to a violet; but if a portion of the water be evaporated, and the remainder be fuffered to fettle as above, no alteration will happen, unlefs the water contain alfo Gypfum, or fome alkali.

c. On adding the acid of fugar to this water a precipitate takes place, as defcribed (page 62. b.) it combines with the calcareous earth, and falls to the bottom, while the fixed air remains diffolved in the water.

d. A folution of foap being added to this water, the fame appearances take place as defcribed (page 56. e.) The lime and oil forming an earthy foap, while the fixed air unites to the alkali.

B. e. The fixed air contained, if in any quantity, may be difcovered by the method defcribed (page 46. d. e.) or by adding a little vitriolic acid to it before the diftillation. The vitriolic acid unites with the lime, and forms Gypfum, which remains diffolved in the water, while the fixed air is detached.

ALUM.

Vitriolic Acid combined with argillaceous Earth.

Synonymous Names.

ALUM. { Argillaceous vitriol.
Aluminous ſulphate.
Sulphate of alumine.

Alum has a very aſtringent and acid taſte; its cryſtals are of the figure of an octohedron; it is very rarely found in mineral waters.

A. a. The earthy baſis may be detected by the addition of lime water, which occaſions a turbid appearance, and a flaky precipitate will be gradually formed, which is the earth of alum. The vitriolic acid, uniting with the calcareous earth of the lime water, forms Gypſum, which is more ſoluble in water.

b. A ſolution of alkali, either cauſtic or mild, vegetable, foſſil, or volatile, added to aluminous water, produces a white flaky appearance. The alkali combines in the ſame manner with the vitriolic acid, and precipitates the earth of alum, either alone or combined with fixed air, leaving vitriolated tartar, Glauber's ſalts, or volatile ammoniac, according to the alkali made uſe of, diſſolved in the water.

c. By adding a ſolution of ſoap, the aluminous water becomes turbid, and depoſites a number of white flakes, the alkali of the ſoap uniting

ing with the vitriolic acid, and the unguinous part with the earth, forming an earthy ſoap.

B. d. The vitriolic acid may be detected by the addition of a ſolution of muriated ponderous earth, white clouds will be inſtantly formed, and a white inſoluble precipitate, which is an artificial ponderous ſpar, will be gradually depoſited at the bottom of the veſſel, as related in other inſtances, where vitriolic acid conſtitutes one of the elements, (page 49, a.)

e. By adding the ſolution of nitre of ſilver, nitre of mercury, or ſugar of lead, to water containing alum, the ſame appearances will take place as enumerated (page 49, b. page 57, h. i. k.) the aluminous earth uniting with the acids, in the ſame manner as the magneſia in thoſe inſtances.

f. A little ſolution of blue vitriol being added, a green precipitate will be formed, with a ſmall tinge of blue in it.

CLASS

CLASS III.

METALLIC SUBSTANCES.

AËRATED IRON,

Iron combined with fixed Air.

	Synonymous Names.
AËRATED IRON.	Aperient ſaffron of mars.
	Ruſt of iron.
	Martial chalk.
	Martial mephite.
	Carbonate of iron.

Both iron and fixed air being found in almoſt every ſituation, and readily uniting together, and, eſpecially where the fixed air is in any quantity, readily diſſolving in water; it is no wonder that this is among the moſt common ſubſtances met with in ſprings. Waters containing it are called chalybeate, or ferruginous waters; and, from their ſtrengthening medicinal qualities, are very generally uſed. They have ſomewhat of a ſtyptic or irony taſte, mixed with ſome pungency, from the ſuperabundant fixed air they contain.

This water is very eaſily known.

A. a. In being expoſed to the air, or heated, it precipitates a brown powder, which is calx of iron. The fixed air being decompoſed; the water will have a rainbow-coloured pellicle on it.

(1) b.

(1) b. On adding the tincture of galls, a purple-coloured precipitate will appear, which, if in any quantity, will be almost black; but if the fixed air be first suffered to evaporate, and the iron to precipitate as above, the liquor will effect no change with the tincture of galls.

(2) c. On adding the smallest quantity of Prussian lixivium, a beautiful blue colour will be precipitated. The Prussic acid, uniting with the iron, forms Prussian blue; neither will this happen, if the precipitation a. be suffered to take place first.

d. On adding a little solution of soap, the water will become turbid, and white flakes will be precipitated to the bottom of the vessel; the alkali of the soap uniting to the fixed air, and the iron to the oil, forming a metallic soap.

e. On adding a solution of volatile liver of sulphur, a precipitate of a very dark green colour is formed. The sulphur combining with the iron, becomes insoluble in the water.

B. f. The fixed air may be detected as described (page 46, a. e.)

(1) *Tincture of galls is made by infusing powdered galls in spirits of wine.*

(2) *Prussian lixivium is formed by digesting Prussian blue with caustic vegetable alkali and water, and then adding a little distilled vinegar to neutralize any superabundant alkali, if alkali should prevail.*

VITRIOLATED IRON.

Iron combined with Vitriolic Acid

Synonymous Names.

VITRIOLATED IRON.
- Sal Martis.
- Copperas, or green copperas.
- Vitriolum viride, green vitriol.
- Martial vitriol.
- Vitriol of iron.
- Sulphate of iron.

Although this ſalt is formed from the decompoſition of martial pyrites, which decompoſe very readily, and are extremely common; and is formed alſo by other natural proceſſes; and although this ſalt is by no means uncommonly found ſolid, in it's native ſtate, yet it is very rarely met with in mineral waters. Its chryſtals are of a green colour.

It is detected nearly as the former.

A. a. By expoſing it to the air it will not ſo ſoon, or ſo perfectly, depoſite the iron in a brown precipitate as the former page 71. The vitriolic acid is not ſo readily decompoſed; it will have the ſame rainbow coloured pelicle on it.

b. On adding the tincture of galls, the ſame purple colour will appear as deſcribed p. 72,

c. On adding the Pruſſian lixivium the ſame colour will be formed, as page 72, and that although it has been firſt heated.

d. On adding a ſolution of cauſtic alkali the liquor will become yellow, and, after ſome time, depofite an ochry fediment. The alkali, uniting with the vitriolic acid, depofites the iron, which unites with a portion of pure air, forming a calx of iron.

e. On adding a folution of foap, the fame appearances take place as defcribed (page 72, d.)

f. On adding a folution of liver of fulphur, a black precipitate will immediately appear: if to this black precipitate a few drops of vitriolic acid be added, it will be perfectly rediffolved as before.

g. If water be added, impregnated with hepatic air, the fame effect will be produced.

B. h. The nature of the acid may be more perfectly determined, by adding a folution of muriated ponderous earth, when white clouds will inftantly be formed. The ponderous earth, uniting with the vitriolic acid, forms barofelenite, which is precipitated, while the iron, combining with the muriatic acid, remains diffolved in the water; for, upon adding the Pruffian lixivium, or tincture of galls, the blue or purple precipitates will appear,

i. By adding a folution of nitre of mercury, a beautiful yellow precipitate will be obtained. The vitriolic acid uniting with the mercury forms turpeth mineral, while the nitrous acid, uniting with the iron, remains diffolved in the water, as may be known by adding the tincture of galls, or Pruffian lixivium.

VITRI-

VITRIOLATED COPPER.

Copper combined with Vitriolic Acid.

Synonymous Names.

VITRIOLATED COPPER.
- Blue vitriol.
- Blue copperas.
- Vitriol of cyprus.
- Vitriol of copper.
- Vitriol of venus.
- Sulphate of copper.

Its chryſtals are of a beautiful blue colour, of an oblong rhomboidal ſhape, the taſte is ſtyptic, even cauſtic. This is often found in mineral waters, in the neighbourhood of copper mines.

A. a. It is detected. By adding lime water, when a pale green precipitate will appear, which, upon pouring in more lime water, is rediſſolved, the liquor becoming a ſapphire blue.

b. By adding the cauſtic volatile alkali, the water alſo becomes of a ſapphire blue colour. The volatile alkali uniting with the copper, forms cuprum ammoniacum, which will remain diſſolved in the water, unleſs it be evaporated.

c. But if the mild volatile, mineral, or vegetable alkali be employed, green clouds will be formed, and in time a precipitate of that colour will fall to the bottom. Theſe alkalis unite with the vitriolic acid, while the copper unites with the fixed air of the alkalis, forming what is called mountain copper, which is precipitated; but if much mild volatile alkali be added, it will rediſſolve the precipitate, and form cuprum ammoniacum.

d. Upon adding a ſolution of Epſom ſalt, or of nitrous or muriated magneſia, the precipitate will be of a pale green colour.

e. Upon adding a ſolution of nitrated or muriated calcarious earth, the precipitate will be ſtill of a more pale green colour.

f. On adding a ſolution of Alum, a green precipitate will be formed, with a tinge of blue in it. In theſe three inſtances, the earths unite with the acid, and the copper is precipitated.

g. If a ſolution of arſenic in water be added, clouds of yellowiſh green colour will be formed, and a ſubſtance of the ſame colour will be precipitated. This is called ſcheel's green.

h. On adding a ſolution of volatile liver of ſulphur, clouds of a blackiſh brown hue will be formed, and a precipitate of the ſame colour fall to the bottom;

i. Or if water, impregnated with hepatic air, be added, the ſame effect will take place.

k. On adding a little of the Pruſſian lixivium, a precipitate of a brown colour will be depoſited.

B. 1. The nature of the acid may be detected by adding a ſolution of muriated ponderous earth, nitre of ſilver, and the other means mentioned (page 49, a. b. &c.)

VITRIOLATED ZINC.

Zinc combined with vitriolic Acid.

Synonymous Names.

VITRIOLATED ZINC.	White vitriol. White copperas. Sal vitrioli. Vitriol of zinc. Goslard vitriol. Sulphate of Zinc.

This is sometimes found in waters near the mines of zinc; it has a styptic taste; its chrystals are colourless, and generally tetrahedral prisms, terminating in pyramids.

a. By adding limewater pearl coloured clouds will appear, and a precipitate of this kind will fall to the bottom.

b. Vegetable alkali, mild or caustic, being added, a precipitate will appear of a dirty yellowish pearl colour.

c. Upon adding caustic volatile alkali, a pale orange coloured preciptate will be formed.

d. But if the volatile alkali be employed in its mild state, a precipitate, nearly of a white colour will be formed. These several substances unite with the vitriolic acid, and remain dissolved in the water, whilst the zinc is precipitated in the form of a calx.

e. Upon adding the Pruffian lixivium, a dirty white precipitate is formed, which becomes yellow in the fire, and becomes white again, on cooling. The Pruffian lixivium will not precipitate earths from folutions of earthy falts in water.

f. Liver of fulphur, or hepatic air, precipitates zinc of a pale coffee colour.

g. Moft of the precipitates of zinc will be found to be volatile in the fire.

ARSENIC.

Regulus of Arfenic combined with pure Air.

Synonymous Names.

ARSENIC. { White arfenic.
Calx of arfenic.
Flowers of arfenic.
Lime of arfenic.
Oxide of arfenic.
White oxide of arfenic.
Sublimed white oxide of arfenic.
Oxygenated arfenic. }

Arfenic is found in fpring waters in the neighbourhood of mines; and as fuch waters are extremely poifonous, it is important to detect it. Arfenic is very foluble in water, on evaporating the water it cryftallizes into triangular pyramida cryftals of a yellowifh colour.

a. If

a. If it exifts in any quantity in water, it may be obtained by fimple evaporation, and either cryftallizing the arfenic or evaporating to drynefs, and then expofing the refiduum to a red heat, in which it will fly off, giving out a ftrong fmell of garlic.

b. On adding a folution of liver of fulphur, or hepatic air, to a folution of arfenic in water, yellow clouds will be produced, and a precipitate of the fame colour will fall to the bottom. The arfenic uniting with the fulphur, forms orpiment, which is precipitated.

c. A folution of blue vitriol being added, clouds of a yellowifh green colour will be formed, which gradually fall to the bottom. The arfenic, uniting with the copper, forms what is called fcheel's green.

d. On adding a folution of gold to water containing arfenic, the gold, after fome time, will be precipitated in its metallic ftate.

e. Upon adding a folution of vitriolated iron, a precipitate, of a dirty green colour, will fall to the bottom. The iron combining with the arfenic.

f. Moft of thefe precipitates alfo will be found volatile in heat, giving the fmell of garlic as they fly off.

It has been the cuftom to give mucilaginous drinks to perfons that had fwallowed arfenic. Milk, fat, oil, butter, &c. But Navier, a Phyfician at Chalones, in France, recommends, from his own experience, the calcarious, or alkaline, livers of fulphur as the beft means of counteracting this poifon. Thefe combine with arfenic in the humid, way faturate it, and almoft diveft it it totally of its causticity. They act still better, if impregnated with a little iron. He prescribes a dracm of these livers of sulphur dissolved in a pint of water, to be drank at various draughts, or five or six grains of them to be made into pills; drinking a glass of warm water after each dose: and he recommends the use of sulphureous waters for some time after the first symptoms are removed. He assures us, that he has seen the happiest effects of this in removing the tremors and paralytic affections, occasioned by this poison.

CLASS

SECT. V.

Of the Contents of the Air.

Did not the acid vigor of the mine,
Roll'd from ſo many thund'ring chimneys, tame
The putrid ſteams that overſwarm the ſky;
This cauſtic venom would perhaps corrode
Thoſe tender cells that draw the vital air,
In vain with all their unctuous rills bedew'd *(d)*.

THAT thin, tranſparent, inviſible fluid, called the atmoſphere, which ſurrounds the earth, riſes to a conſiderable height above its ſurface. Although inviſible to us, it is a real ſubſtance, excluding other ſubſtances from the ſpace which it occupies, and, like other fluids, preſſes equally in all directions: its weight is the cauſe of the ſuſpenſion of mercury in the barometer, and is to that of water as 1 to 900, or 15 pounds upon every ſquare inch of the earth.

It chiefly conſiſts of two elaſtic fluids, or gaſes *(e)*, poſſeſſed of very different and oppoſite properties; it contains alſo other gaſes, or ſubſtances, ſuſpended or diſſolved in it, but in ſmall quantity. Theſe two fluids, or gaſes, are diſtin-

(d) Armſtrong's Art of preſerving Health, p. 51.

(e) Gas, from gaſcht (German) an eruption of wind; any matter ſubtilized by heat into an elaſtic aeriform ſtate.

guiſhed by the names of oxygen gas, or the pure part of the atmoſphere, firſt diſcovered by Dr. Prieſtley in the year 1774, and by him called dephlogiſticated air; by others vital air, becauſe it is the only air that is capable of ſuſtaining the vital principle, or of ſupporting the combuſtion of inflammable matter: it is not only the ſole vivifying power, but the parent of acidity, and hence named oxygen *(f)*, from the moſt general property which its baſe poſſeſſes of forming acids, by combining with different ſubſtances.

The other, or noxious portion of the atmoſphere, is termed azote *(g)*, from its known quality of killing animals; hence the noxious part of the atmoſpheric air, which is totally unfit for reſpiration, is called azotic gas, or phlogiſticated air.

The atmoſphere contains other elaſtic fluids analogous to common air, with reſpect to elaſticity and inviſibility; but otherwiſe eſſentially different from each other; ſuch as fixed air, or carbonic acid gas; inflammable air, or hydrogen gas; nitrous gas: beſides a variety of other ſubſtances, reſulting from various exhalations conſtantly emitted from vegetable, animal, and foſſil bodies *(h)*; but it is to the gaſes above enumerated that the following obſervations will be confined.

Atmospherical

(f) Oxygen, οξυς, acidum, or acid, and γινομαι, gignor, to produce, or generate acids.

(g) Azotum, αζωτον, from a privation of, and ζων, life.

(h) *Fluoric acid gas*, which is diſengaged from native fluate of lime, or vitreous ſpar, by ſulphuric acid. *Muriatic acid gas*, or muriatic acid purified from water, and melted by caloric into an elaſtic

Atmospherical Air, or common Air.

As the two gafes of oxygen and azote exift in the air, merely in a ftate of intimate mixture, and not in chemical combination, they may be eafily feparated by a fubftance poffeffing an electric attraction for either of them: thus, if pure mercury be expofed for fome time in clofed veffels filled with atmofpheric air, to a heat nearly approaching to ebullition, a quantity of air will difappear, and the mercury be converted into a red powder, oxyd, or calx of mercury, which will have gained an addition of weight equal to that of the air abforbed, the mercury having united with the oxygen gas, which leaves the azote. If a bell glafs be inverted over a folution of liver of fulphur, the air in it will fuffer a diminution; the oxygen will be abforbed, and the unabforbed part will be azote, fhowing that there are 27 parts of oxygen, and 73 parts of azote in 100 parts of atmofpheric air.

If the azote in liver of fulphur be expelled, and the oxygen in the red calx of mercury, or

elaftic fluid. *Oxygenated muriatic acid gas*, or *dephlogifticated marine acid*, which is difengaged during the reciprocal action of native oxyd of manganefe and muriatic acid, being produced by the tranfition of oxygen from the manganefe into the muriatic acid. *Ammoniac gas, alcaline air*, or *volatile alcaline gas*, is difengaged by heat from liquid ammoniac; or, from a mixture of ammoniacal muriate, or common fal ammoniac, with quick lime. *Hepatic gas*, or fölphurated hydrogenous gas, is obtained from folid alcaline fulphures, or livers of fulphur, by decompofing them with acids. *Phofphorated hydrogenous gas*, or *phofphoric gas*, obtained by boiling a lixivium of cauftic pot-afh with half its weight of phofphorus, and receiving the elaftic fluid difengaged into glaffes containing mercury.

 precipitate

precipitate per ſe, and theſe be mixed, atmoſpheric or vital air will be formed.

Oxygen gas conſtitutes about one fourth of the atmoſphere, and is compoſed of light, caloric *(i)*, and oxygen. It is procured, by filling a receiver with the green leaves of vegetables, and inverting it in ſpring water, and expoſing it to the direct rays of the ſun; the leaves yield a conſiderable quantity of oxygen air, which aſcends to the upper part of the receiver, and may be eaſily removed from it for uſe. It is produced from various other ſubſtances, and particularly from nitre, or the metallic calces. One ounce of nitre, expoſed to a red, or rather white heat, in an earthen retort for about four or five hours, will give about 700 cubic inches of oxygen air; as this gas contains a quantity of nitrous acid in the form of vapour, it may be ſeparated from it by agitating the air in lime water.

But it is from manganeſe that the greateſt quantity of this gas may be procured, and it is now the ſubſtance generally uſed. One ounce of good manganeſe will, in a red heat, yield more than two pints and a half, or about 80 cubic inches of elaſtic fluid, one tenth of which is carbonic acid, and the reſt oxygen gas.

(i) Caloric, latent heat, fixed heat, or principle of heat, is only known by its effects; inviſible, imponderable, ſubtile, pervading all bodies, inſinuating itſelf between their particles. It melts ſolid bodies, rarefies fluids, and renders them inviſible. There are but three ſpecies of the undecompounded bodies, which are known to be rendered into the ſtate of gas by coloric; namely, oxygen, hydrogen, and nitragen.

	Synonymous Names.
AZOTIC GAS,	Vitiated air. Impure air. Phlogisticated air. Phlogisticated gas. Atmospheric mephitis.

Is composed of caloric, or heat, and a particular base, capable of becoming solid, called azote. This substance, united to different bases, forms alkalies, and may be considered as a real alkalizen, or alkalizing principle, in opposition to oxygen, which is the principle of acidity. The atmosphere, therefore, is an immense reservoir of the principles of acidity and alkalescency, without being itself either acid or alkali.

Azotic gas may be obtained by exposing certain substances to atmospheric air, which absorb its oxygen; as ammonia, or the volatile alkali, which is composed of azote, united with hydrogen.

	Synonymous Names.
CARBONIC ACID GAS,	Fixed air. Solid air of Hali. Cretaceous acid gas. Mephitic gas. Aerial acid.

Is the heaviest of the aerial fluids, being composed of carbon or charcoal, and vital air or oxygen, forming carbonic acid. It is combined with calcareous

careous earth or lime, to which it has a greater affinity than to any other; it conſtitutes nearly half the weight of chalk, limeſtone, or carbonate of lime, &c. converting them into ſaline ſubſtances, or carbonates; it is produced by burning limeſtone, in which the carbonic acid combines with caloric, and flies off in the form of gas, leaving the calcareous earth pure, or in a ſtate of quicklime; it is produced plentifully by fermenting liquors, and is often found in coal mines, called the chalk, or chalk damp: it is formed in caverns, as in the Grotto del Cano, which extinguiſhes flame and deſtroys life. Charcoal, on burning, unites with the oxygen of the atmoſphere, and forms carbonic acid, which is alſo deſtructive to life, and extinguiſhes the light of a candle.

Water abſorbs more than its bulk of this gas, and acquires an acid taſte and ſparkling appearance, and moſt mineral waters owe their ſparkling ſpirituous taſte to this gas; and cyder, beer, and other fermented liquors owe their brightneſs to the carbonic acid which they contain. Water thus impregnated is capable of diſſolving a ſmall quantity of iron, forming a chalybeate water like Spa, or Pyrmont. When fixed alkali is previouſly diſſolved in water, it will then abſorb a greater quantity of carbonic acid than common water, and forms the aqua mephitica alcalina. Limeſtone, or carbonat of lime, cannot be diſſolved but in a very ſmall quantity in water; but pure, or quicklime, can be diſſolved in a conſiderably greater proportion, forming lime-water, which is a teſt of carbonic acid; for it is certainly

precipitated

precipitated by that acid in the form of carbonate of lime, or limeſtone. If the water be ſaturated with carbonic acid, it will diſſolve carbonate of lime in conſiderable quantity; and if this ſolution be let fall drop by drop on any ſubſtance, or the carbonic acid eſcapes, the carbonate of lime will be depoſited in the incruſted ſubſtance with calcareous earth, and the ſubſtance is ſaid to be petrified.

HYDROGEN*(k)* GAS,

OR

INFLAMMABLE AIR,

Inflammable Gas, or Phlogiston of Kirwan.

Hydrogen, which is the baſe of water, combined with caloric, forms hydrogen gas, or inflammable air, diſcovered by Mr. Cavendiſh in 1767. It is the lighteſt of elaſtic fluids, being twelve times lighter than common air; hence employed in air-balloons. It inflames by contact of an ignited body, but will burn only in contact with common, or oxygen, air. It is leſs noxious than the carbonic acid: ſuffers no diminution on mixture with nitrous air. It is produced during the diſſolution of animal and vegetable bodies; hence it

(*k*) Hydrogenium (υδρογενιον, from υδωρ, water, and γινομαι, to become, or γενναω, to produce) hydrogen, one of the principles of water. The baſe of that elaſtic fluid, which was formerly called inflammable air.

is

is found to come out of ponds, burying-grounds, and other places containing animal and vegetable matter in a ſtate of decay. Found alſo in the earth where inflammable minerals are contained; when, combining with atmoſpheric air, it ſometimes ſuddenly takes fire to the danger of the miners, and by them called fire-damps. It is obtained from moſt kind of bodies; but the greateſt quantity may be extracted from zinc, by means of diluted vitriolic acid; and from red-hot iron, by paſſing the ſteam of boiling water over its ſurface. In like manner charcoal yields inflammable gas, called hydrocarbonate, combined with the carbonic acid gas. It is this aeriform fluid which floats frequently on marſhes, and being ſet on fire by electricity, or other means, gives riſe to the ignis fatuus. Hydrogen gas has the property of diſſolving and ſuſpending a variety of ſubſtances, as iron, charcoal, ſulphur, phoſphorus, &c. hence we have the names of phoſphoric hydrogen gas, or phoſphuret of hydrogen; ſulphuric hydrogen gas, or ſulphuret of hydrogen; with phoſphorus it is called phoſphorated hydrogen gas, giving out a fœtid ſmell, is improper for reſpiration, and takes fire ſpontaneouſly in coming in contact with the air, accompanied with an exploſion; and probably to the diſengagement of gas, the ignis fatuus, playing about burying-grounds and places where animals are putrifying, may be attributed.

Nitrous acid gas is diſengaged from nitric acid by various combuſtible bodies, eſpecially metals, oils, mucillages, and alcohol. By the action of the nitric acid on theſe ſubſtances it becomes

comes nitrous gas; it extinguishes light and de stroys animals, being totally unfit for animal respiration and combustion. It is not dissolved in water, nor does it indicate the least property of an acid: by combination with vital air it affords nitric acid, being itself nothing but nitric acid deprived of a part of its oxygen, and consequently a compound of azote, or nitrogen and oxygen, containing, however, more azote and less oxygen than the nitric acid; hence are produced the varieties of this gas, according to the different proportions of azote and vital air. In nitrous gas the azote and oxygen are deprived of all that quantity of caloric and light which they possessed in the atmosphere: the oxygen, however, still retains enough of both these principles to occasion a combustion, with flame, of several combustible bodies when immersed in it, as a pyrophorus, or a small portion of sulphuret of alkali; the nitrous gas is then gradually diminished, and at length pure nitrogen gas remains. In this case, the oxygen of the nitrous gas combines with the inflammable body, and thus the nitrogen gas is left in a free state; hence, as has been intimated, it is evident that azote, or nitrogen, constitutes the basis of nitric acid, which obtains its acid property from its combination with oxygen. The mixture of one part of nitrous gas, and four parts of oxygen gas, produces red vapours; because the bases of each gas mutually combine, and form nitrous acid gas; and the superabundant caloric escapes in the state of sensible heat. If this operation be instituted over mercury, the nitric acid recently formed will remain upon its surface, in the form of nitrous acid gas; but if made over water, it is immediately abſorbed by it.

As azotic gas does not combine with nitrous gas, but remains behind when mixed with it, Dr. Prieſtley employed nitrous gas for aſcertaining the proportionate quantity of oxygen gas in atmoſpheric air. To this end equal parts of nitrous gas and atmoſpheric air are mixed over water, and the greater or leſs diminution of volume of gaſes determines the relative quality of the air examined.

To aſcertain the purity of atmoſpheric or other airs, different kinds of eudiometers have been deſcribed by Prieſtley, Magellan, and Abba Fontana. A ſimple method is deſcribed by Dr. Archer, which is, to have a tube, between two and three feet in length, one third of an inch wide, and open at one end, a meaſure for the airs; a ſcale carefully graduated, according to the contents of the meaſure, into tenths and hundredth parts, ſo that one of the hundredth parts will be about one-ſixth of an inch. The operator is to fill the tube with water, and invert it in a trough of the ſame fluid, and ſet it over a hole on the ſhelf of a trough; the tube will remain perfectly full of water, from the preſſure of the atmoſphere upon the ſurface of the water in the trough. Then fill the air meaſure with water alſo, and if it be the air of the place to be tried, it is only neceſſary to pour out the water, and the air will occupy the ſame ſpace in the meaſure as had before been occupied by the water. Then convey the air, contained in the meaſure, under the ſurface of the water in the trough, to the bottom of the tube; by gently turning

turning the hand, the air in the meaſure may be made to aſcend into the tube, within which it will cauſe a portion of the water to deſcend, exactly correſponding to the quantity of air that was contained in the meaſure. Now mark the ſpace which the air occupies, by applying the graduated ſcale to the outſide of the tube; next take the ſame meaſure full of nitrous air, and introduce it alſo in the ſame manner into the tube; gently agitate the tube, and, after waiting ſome little time, apply the ſcale again, and obſerve what diminution has taken place; for if two meaſures of the ſame air had been introduced, they would occupy of courſe twice as much ſpace within the tube as one would have done. But this will by no means be the caſe in the inſtance before us; for a meaſure of common air, and one of nitrous air, will not occupy as much ſpace as two meaſures of common air, or two of nitrous would, but will fall conſiderably ſhort of it; and it is in proportion to this diminution that the purity of the air is aſcertained, the greater diminution denoting the greater purity of the air. Thus, if one meaſure of nitrous air be added to one of any other kind of air, and the diminution be $1\frac{26}{100}$, the meaſure of the teſt will be 1.26. The nitrous acid being compoſed of nitrous air and oxygen air, whenever theſe two airs meet, they put off their aëriform ſtate, and become liquid nitrous air; which, as it mixes with the water, no longer impedes the aſcent of that fluid in the tube.

The ſpecific gravity of different kinds of air:

A cubic inch of atmoſpheric air weighs		385.
dephlogiſticated	-	420.
phlogiſticated	-	377.
fixed -	-	570.
inflammable	-	35.
nitrous -	-	399.

SECT.

SECT VI.

Directions for collecting and distinguishing Fossil *substances, including* Salts, Earths, Metals, *and* Inflammables.

TO write particularly upon these objects of natural history is not the present design, but to give such general instructions, as may assist a traveller, in the choice of fossil bodies, till fuller information can be procured, which the numerous modern authors on these subjects afford. In the arrangement adopted, I have principally had in view that by Dr. Babington; and his own and the other references annexed to the descriptive histories, will point out the principal writers on the subjects of discussion.

The traveller should be furnished with flint and steel, and the mineral acids; at least with the nitrous, and volatile acids; a hammer also will be necessary, to break bodies too bulky to bring away.

A blow-pipe is likewise an useful article; by blowing long and forcibly through such an instrument, upon the flame of a candle, by which the point of the flame may be directed against the body to be examined, it will frequently discover whether it is a calcareous, vitrifiable, or refractory substance; and for greater precision, experiments

ments for the ſame end may be afterwards made at the fire-ſide, on a charcoal fire. Even a tobacco-pipe may be uſed as a crucible in a common fire.

The collector ſhould alſo attend to as many of the following particulars as poſſible.

1. When any article is collected, mark it by a number, or ſome ſign of diſtinction, referring to a catalogue, with all the particulars that may be known relative to it; as,

2. Where it was found.

3. In what quantity, whether ſcarce or abundant.

4. If on the ſurface of the earth, or at what depth.

5. In what poſition, whether horizontal, perpendicular, &c. And with what other foſſil bodies it was found; as clay, ſtone, ſlate, mineral, &c.

6. Whether in ſtrata, or looſe nodules.

7. The depth and thickneſs of the ſtrata, how they incline, or to what points of the compaſs they tend; or if level or horizontal; whether they have perpendicular or horizontal fiſſures, and what foſſil bodies are contained in theſe fiſſures.

8. All high mountains and hills, eſpecially their ſides, are to be ſearched; the ſhores alſo of the ſea, with their banks, and the cliffs adjacent, and the falls of caſcades, rivers, and great gullies.

9. The ſituation of mines, pits, and quarries, whether in a valley or hill; and the diſpoſition of the ſtrata, whether horizontal, inclining, &c. their thickneſs, and the depth they lie; and what other foſſils are imbedded in the ſtrata, or in the neighbouring caverns, fiſſures, partings, &c.

10. The

10. The waters of mines ſhould be examined, whether pure, taſteleſs, purgative, vitriolic, or chalybeate, &c.

11. The damps and ſteams of mines, and what are the conſequences or effects of them; in what ſeaſons and ſtate of the air they are chiefly obſerved; and what temperature the air bears in different depths of ſuch mines.

12. The account given by the natives, inhabitants, miners, workmen, &c. who may be acquainted with the ſubject.

CLASS I.

SALTS.

1. Simple Salts,

Having radicals, which are simple and known.

A. *ACID.* All acids reſemble one another in taſte; in their manner of giving a red colour to vegetable ſubſtances; in their common tendency to combine with earths, alkalis, and metallic oxyds.

Synonymous Names.

a. *Carbonic.*
- Gas ſylveſtre.
- Spiritus ſylveſtris.
- Fixed air.
- Aërial acid.
- Atmoſpheric acid.
- Mephitic acid.
- Cretaceous acid.
- Carbonaceous acid.

See Sect. V. page 45, 85.
Anal. Carbon. 17. Oxygen 83. *Chaptal.*

Synonymous Names.

b. *Boracic.*
- Volatile narcotic ſalt of vitriol.
- Sedative ſalt.
- Acid of borax.
- Boracine acid.

This

This acid is dry, cryſtallized in hexahædral plates, almoſt inſipid, ſcarcely ſoluble, fuſible with ſilices into a glaſs, very feeble in its affinities, and liable to yield up the terrene and alcaline baſes to almoſt any other acid.

	Synonymous Names.
c. *Sulphuric.*	Acid of ſulphur. Vitriolic acid. Oil of vitriol. Spirit of vitriol.
Sulphureous.	Sulphureous acid. Volatile ſulphureous acid. Phlogiſticated vitriolic acid. Spirit of ſulphur.

The *ſulphuric acid* is formed from ſulphur and oxygen, by the combuſtion of the ſulphur; inodorate, twice as heavy as water, and leſs volatile, very cauſtic; affording ſulphureous acid gas and ſulphur, by its decompoſition with *red hot coal*, metals, &c. forming ſulphates*(l)* with the earths, alkalis, and metallic oxydes.

Anal. Sulph. 72. Oxygen 28. *Berthollet.*

Sulphureous acid is highly odorate, volatile, gaſeous, deſtructive to blue vegetable colours, and fit to cleanſe away ſpots made with theſe colours on a white ground; abſorbing, by ſlow degrees, the oxygen from the air, and from various

(l) Salts formed by the combination of the ſulphuric acid with different baſes, decompoſable by charcoal, &c. into ſulphures.

acids and oxydes; forming ſulphites (m) with the earthy and alkaline baſes.

2. Compound Salts.

A. *Base POT-ASH*, or *Vegetable, caustic, fixed alkali.* The impure ſalt, containing alkali, which is obtained by evaporating the aqueous ſolution of wood aſhes to dryneſs; it is then ſolid, white, cryſtallized in rhomboidal plates.

	Synonymous Names.
a. *Carbonate** *of Pot-ash*, or pot-aſh combined with carbonic acid;	Fixed ſalt of tartar. Vegetable fixed alkali. Aërated vegetable fixed alkali. Cretaceous tartar. Mephitic tartar. Mephite of Pot-aſh. Nitre fixed by itſelf. Alkaheſt of Van Helmont.

It forms regular cryſtals, which repreſent tetrahædral priſms, terminating in ſhort tetrahædral points; has an urinous taſte, and changes the ſyrup of violets green.—*Anal.* Acid 23. Alkali 70. Water 5. *Bergman.*

(m) Salts formed by the combination of the ſulphureous acid with different baſes, and yield by the contact of almoſt any acid, the ſmell of burning ſulphur with effervescence.

* A ſalt formed by the union of carbonic acid with different baſes.

b. *Muriate*

b. *Muriate** *of Pot-ash*, or Febrifuge ſalt of Sylvius; compoſed of muriatic acid and pot-aſh; is found in its native ſtate in ſea-water; in the mother earth of nitre manufactories; in various ſalt-ſprings; has rhomboidal or octahædral cryſtals.—*Anal.* Acid 31. Alkali 60. Water 8. *Bergman.*

c. *Nitrate*† *of Pot-ash, Nitre,* or *Saltpetre;* formed by the combination of nitric acid and pot-aſh. It is found in the fiſſures of lime-ſtone kilns, near Molfetta, in the kingdom of Naples; in various waters, and even in rain; and in the freſh juices of many plants.—*Anal.* Acid 33. Alkali 49. Water 18. *Bergman.*

	Synonymous Names.
d. *Sulphate*‡ *of Pot ash.*	Vitriol of pot-aſh.
	Sal de duolus.
	Vitriolated tartar.
	Arcanum duplicatum.
	Sal polychreſt of Glaſer.

compoſed of the ſulphuric acid and pot-aſh: ſcarcely ever found native.—*Anal.* Acid 40. Alkali 52. Water 8. *Bergman.*

* Salts formed by the union of the muriatic acid with different baſes.

† Salts formed by the combination of the nitric acid with different baſes.

‡ Salts formed by the combination of the ſulphuric acid with different baſes.

B. *Base SODA.* Cauſtic ſoda, marine alkali, or mineral alkali united to carbonic acid, which is obtained from the aſhes of the kali ſpinoſum, and of many ſea plants.

Synonymous Names.

a. *Carbonate of Soda,*
- Natrum or natron.
- Baſe of marine ſalt.
- Marine or mineral alkali.
- Cryſtals of ſoda.
- Cretaceous ſoda.
- Aërated ſoda.
- Effervеſcent ſoda.
- Mephite of ſoda.
- Aërated mineral fixed alkali.
- Effervеſcent mineral fixed alkali.
- Chalk of ſoda.

is a perfect neutral ſalt, conſiſting of the carbonic acid, and pure or cauſtic ſoda; has rhomboidal cryſtals. Found in Hungary, Egypt, Perſia, the Eaſt Indies, and China; is found ready formed on the ſurface of the earth.—*Anal.* Acid 16. Alkali 20. Water 47. *Delamethrie.*

b. *Borate* of Soda.* Greyiſh and greeniſh white, moſtly in tabular, ſix-ſided priſms. Found on the ſnow mountains of Tibet, &c.—*Anal.* Acid. 34. Alkali 17. Water 47. *Delamethrie.*

c. *Muriate of Soda,* or marine ſalt, or culinary ſalt. Cryſtallized in cubes, amorphous,

* Borax Tinkal.

fibrous

fibrous, compact. *Anal.* Acid 52. Alkali 42. Water 6. *Bergman.*

d. *Sulphate of Soda.* Glauber salt, or vitriol of soda. Sulphuric acid with soda.—*Anal.* Acid 27. Alkali 15. Water 58. *Bergman.*

c. *Nitrate of Soda.* Cubic nitre, or rhomboidal nitre formed by the combination of the nitric acid and soda.

	Synonymous Names.
C. *Base*, AMMONIAC,	Caustic volatile alkali. Fluor volatile alkali. Volatile spirit of sal ammoniac.
a. *Carbonate of Ammoniac*,	Ammoniacal chalk. Cretaceous ammoniacal salt. Concrete volatile alkali. Ammoniacal mephite. English sal volatile.

Sal ammonia united to carbonic acid.—*Anal.* Acid 45. Alkali 43. Water 12. *Bergman.*

b. *Muriate of Ammoniac*, salmiac, sal ammoniac, or muriated ammoniac, composed of the muriatic acid and ammoniac, or the volatile alkali. It is found ready formed near the craters of volcanos, and in Tartary. In Egypt it is prepared in great abundance by subliming the soot, obtained from the combustion of animal excrements.—*Anal.* Acid 52. Alkali 40. Water 8. *Delametherie.*

c *Nitrate*

c. *Nitrate of Ammoniac*; ammoniacal nitrous ſalt, or ammoniacal nitre; a combination of the nitric acid and ammoniac. It is generated ſpontaneouſly in the mother earth, or ſaltpetre manufactories, though but in ſmall quantities; mixed with nitrate of pot-aſh.—*Anal.* Acid 46. Alkali 40. Water 14. *Delamethrie.*

d. *Sulphate of Ammoniac*; ammoniacal vitriolic ſalt, ammoniacal ſalt (ſecret of Glauber's) or ammoniacal vitriol, formed by the ſulphuric acid and ammoniac; concrete.—*Anal.* Acid 42. Alkali 40. Water 18. *Delamethrie.*

CLASS

CLASS II.

EARTHS.

EARTHS are ſuch bodies as are inodorous, inſipid, pulverulent or friable, not flammable, ſcarcely ſoſuble in water, unchangeable in the fire; when heated, either by themſelves or with inflammable bodies, do not acquire metallic ſplendour.

I. Homogeneous or simple Earths.

Thoſe earths which have not hitherto been decompounded, are termed elementary, homogeneous or ſimple earths. Of this kind are the following nine. A. Lime or calcareous earth. B. Stranthian earth. C. Baryte or ponderous earth. D. Magneſia or earth of bitter ſalt. E. Alumine or argillaceous earth. F. Silex or ſiliceous earth. G. Adamantine earth. H. Jargon earth. I. Sidnean earth; this laſt, however, is doubtful as an elementary earth.

A. *LIME.*

a. *Pure lime,* cauſtic or pure calcareous earth, in a ſtate of mixture with aërated calcareous earth, has hitherto been found only at Bath, and near the craters of ſome ancient volcanos.

b. *Carbonate*

	Synonymous Names.
b. *Carbonate of Lime,*	Chalk.
	Limeſtone.
	Calcareous mephite.
	Aërated calcareous earth.
	Effervescent calcareous earth.
	Calcareous ſpar.
	Cream of lime.
	Calcareous carbonate.

effervesces with acids, convertible into pure, or quick-lime, by calcination; is then ſoluble in 680 times it's weight of water, forming lime-water, in which action it gives out it's latent heat, cauſing the maſs to become hot.—*Anal.* Carbonic Acid 34. Lime 55. Water 11. *Bergman.* Here may be included ſtalactites, marble, limeſtone, chalk, zufa, ganil.

c. *Swine-stone,* carbonated lime impregnated with petroleum; colour, Iſabella yellow, bluiſh, or blackiſh, or yellowiſh grey, is opaque, compact, and partly of a glittering appearance; when broken, exhibits a ſplintery or conchoidal appearance; when pounded or ſcraped, it emits an urinous or alliaceous ſmell. *Anal.* Carbonate of lime impregnated by petroleum. *Kirwan.*

d. *Sidero-Calcite,* pearl ſpar: colour of a white greyiſh, yellowiſh, or reddiſh white; found amorphous, cellular, rounded, but moſtly cryſtallized; externally metallic or pearly; effervesces with acids. *Anal.* Carbonated lime 60. Oxyde of magneſia, 35. Iron 5. *Woulfe.*

e. *Baryto-*

e. *Baryto-calcite*, of a dark, or light grey, almoſt white; rounded form, or cryſtallized in quadrangular priſms; effervefces with acids. *Anal.* Carbonated lime 92. Carbonated baryte 8. *Bergman.*

f. *Muri-calcite*, Compound ſpar, of an olive colour, clayey appearance; ſtony, amorphous,* and cryſtallized. *Anal.* Carbonated lime 52. Carbonated magneſia 45. Iron and manganeſe 3. *Klaproth.*

g. *Argentine*, Scheifer ſpar, amorphous, foliated. *Anal.* Carbonated lime, with magneſia, argill, and iron. *Kirwan.*

h. *Superſaturated with carbonic acid*, Dolomite, or elaſtic marble, pure or greyiſh white, yellowiſh white, or light red; ſurface rough and uneven; fracture conchoidal or granularly foliated. Effervefces ſlowly with acids; phoſphoreſces when ſtruck in the dark, or when laid on red-hot iron. *Anal.* Lime 44, 29. Carbonic acid 46, 1. Argill 5, 86. Magneſia 1, 4. Iron 0, 074. *Sauſſure, jun.*

Synonymous Names.

i. *Fluate*† *of lime*,
- Fluor ſpar.
- Vitreous ſpar.
- Cubic ſpar.
- Phoſphoric ſpar.
- Sparry fluor.

Nearly inſoluble in water, does not efferveſce with acids, nor ſtrike fire with ſteel; is more or leſs tranſparent, and exhibits various colours; it's

* Amorphous, ex α prriv. et μοςφη, forma.

† A ſalt formed by the combination of the fluoric acid with different baſes.

P ſurface,

ſurface, when ſcratched with a knife, appears greaſy; it generally crackles and phoſphorates by heat, except the colourleſs, which becomes electric when rubbed; and the coloured kinds loſe their colour by heat; but by a ſtrong heat is melted. *Anal.* Fluor acid 16. Lime 57. Water 27. *Scheele.* Oxyde of manganeſe 3. Oxyde of iron 1. Pot-aſh 18. *Vauquelin.*

Synonymous Names.

k. *Phoſphate of lime.* { Earth of bones.
Calcareous phoſphate.
Animal earth.

Apatite, colour various, olive green, bluiſh, yellowiſh and chocolate brown: found cryſtallized in three and ſix-ſided priſms; in ſix and eight-ſided plates, variouſly modified, ſhining and of a greaſy luſtre, texture laminated, ſemi-tranſparent. *Anal.* Phoſphoric acid; lime 55. *Klaproth.*

Synonymous Names.

l. *Sulphate of lime,* { Vitriol of lime.
Calcareous vitriol.
Selenite.
Gypſum.
Plaſter of Paris.

does not effervelce with acids; found in moſt parts of the earth and in water; ſoluble in about 500 times it's weight of water; it reſembles the calcareous ſpar, and is found in ſolid maſſes, of particular form, cryſtallized into hexagonal* priſms, terminating in an edge, or in rhomboids, or cuneform, tabular or lenticular; texture fi-

* Hexagonal, ex ἑξ, ſex; et γωνια, angulus.

brous

brous, or laminated. *Anal.* Sulphuric acid 16. Lime 57. Water 27. *Scheele.*

B. *STRONTIAN*, or Scottish earth, so called from Strontian in Scotland; never found pure. Sulphate of Strontian is said to have been found in great quantities in the neighbourhood of Bristol. At Ham-green also, a variety of this rare production is found breaking through the soil, in such large masses, that it has been made use of in mending the roads.

a. *Carbonate of Strontian, strontianite*; hitherto found in a mild state united to fixed air; in solid masses, of a fibrous texture, striated. Like limestone, it loses it's air in a strong heat, and then forms a lime, heavier than common lime, and more soluble in water. This lime-water has the singular property of affording, when saturated, compressed rhomboidal crystals, which are nothing but pure lime. *Anal.* Carbonic acid 26, 5. Stronthian 73,5. *Kirwan.*

C. *BARYTE*, Ponderous earth; never found in a separate or pure state, but with the sulphuric acid, in the state of ponderous spar or sulphate of baryte; or with the carbonic acid, carbonate of baryte, or mixed with other earths.

	Synonymous Names.
a *Carbonate of Baryte*,	Barotic or ponderous chalk.
	Aërated ponderous earth.
	Effervescent barotes.
	Barotic mephite.
	Barolite.

 found

found in ſolid maſſes, and cryſtallized, hexagonal priſms, terminating in ſix-ſided pyramids; texture ſhining, radiated, fibrous; brittle and ſemi-tranſparent; efferveſces with diluted nitric and muriatic acids, and is entirely diſſolved by them. Expoſed to a ſtrong heat, it parts with it's carbonic acid, and combines with heat, and like lime is ſoluble in water. *Anal.* Acid 20,8. Baryte 78,6. *Withering.*

Synonymous Names.

b. *Sulphate of Baryte,* { Heavy or ponderous ſpar.
Barotic vitriol.
Baroſelite.

found in ſolid maſſes, cryſtallized, approaching to an earthy ſtate, and the tranſparent kind is electric; is four times heavier than water. When expoſed to heat, and to the ſun, it has the property of abſorbing light, which it emits in the dark. Like metals, it is precipitated from it's combinations with the different acids, by the pruſſiate of pot-aſh*. *Anal.* Sulphuric acid 32,8. Baryte 67,2. *Withering.*

Bolognian Stone. *Anal.* Sulphate of baryte 62. Silex 16. Argill 15. Sulphate of lime 6. Water 2. *Arvidson.*

c. *Liver-stone*, of a greyiſh, or greyiſh black colour; texture foliated, ſtriated; does not efferveſce with acids; when heated, or rubbed, it

* Salts formed by the union of the Pruſſic acid, or colouring matter of Pruſſian blue, with different baſes.

emits

emits a ſmell of liver of ſulphur. *Anal.* Sulphate of baryte 38. Silex 33. Sulphate of argill 22. Sulphate of lime 7. Petroleum 3. *Bergman.*

D. *MAGNESIA* is never found naturally pure, but generally combined with the ſulphuric acid, in the form of Epſom ſalt, ſal amarus, or ſulphate of magneſia.

a. *Calci-murite*, amorphous, earthy; colour blue or olive green; of the conſiſtence of clay. *Anal.* Silex 50. With carbonate of magneſia and iron.

b. *Argillo-murite*, of a looſe earthy texture; greeniſh yellow colour, conſiſtence earthy. *Anal.* Magneſia, 13. Silex 50. Argill 10. Lime 3. Oxyde of iron 0,9. Water 13. *Kirwan.*

c. *Silici-murite*, Martial muriatic ſpar, amorphous, foliated. *Anal.* Silex 50. With carbonate of magneſia and iron. *Kirwan.*

d. *Talc*, Talcite, Venetian talc, or Shiſtoſe talc. Generally of a greeniſh or ſilvery white luſtre, like mother of pearl, ſoft and ſemi-tranſparent; texture laminated, foliated, and greaſy, or of a ſoapy feel to the touch; does not efferveſce with acids. *Anal.* Magneſia 44. Silex 50. Argill 6.

e. *Lapis Ollaris*, Pot-ſtone, or Indurated talc; of a pale yellowiſh and greeniſh grey, reddiſh grey, or white colour, with micaceous particles; found in in amorphous maſſes, of a lamellated texture; feels greaſy, does not adhere to the tongue.; becomes hard in the fire. *Anal.* Magneſia 38. Silex 38. Argill 7. Iron 5. Carbonate of lime 1. And a trace of fluoric acid. *Weigleb.*

f. *Steatites*, Fatty ſtone, of a greyiſh, yellowiſh, or greeniſh white, with a ſhade of green colour; found in amorphous maſſes, of an earthy, ſplintery texture, lamellated; the thin lamina ſemi-tranſparent; ſoapy to the touch, does not adhere to the tongue. *Anal.* Magneſia, 20,84. Silex 48,42. Argill 14. Iron 1. Air and Water 16. *Klaproth.*

g. *Serpentine*, of a deep, blackiſh, olive green, ſeldom yellow, ſometimes crimſon red, bluiſh and greeniſh grey; generally exhibits various colours, like the ſkin of a ſerpent; found in ſolid maſſes; when broken has a dull appearance, ſmooth to the touch, and takes a poliſh. *Anal.* Magneſia 35. Silex 45. Magnetic iron 14. Carbonate of lime 6,25. Argill 0,25. *Knock.*

h. *Chlorite*, of a light and dark greeniſh colour, or greeniſh brown; of a ſcaly texture and glittering appearance; feels greaſy or earthy; gives an earthy ſmell, when breathed on. It melts into a dark, black, compact ſlag, and thus becomes magnetic. *Anal.* Magneſia 0,39,47. Silex 0,415. Argill 0,0613. Lime 0,015. Iron 0,1015. Air and water, 0,015. *Hoepfner.*

i. *Asvestos*,

i. *Asbestos*, of a leek or olive green colour, or grey; feels greasy; breaks into parallel, striated, and sometimes curved, fibres or splinters; does not effervesce with acids. *Anal.* Carbonate of magnesia 16. Silex 63,9. Carbonate of lime 12,8. Argill 1,1. Oxyde of iron 6. *Bergman.*

k. *Amianthus*, of a green or greyish, or silvery, white, sometimes olive green; breaks into flexible, and mostly parallel, filaments; soft and greasy, somewhat transparent. *Anal.* Carbonate of magnesia 18,6. Silex 64. Carbonate of lime 6,9. Baroselenite 6. Argill 3,3. Oxyde of iron 12. *Bergman.*

l. *Suber Montanum*, cork-like, or resembling brown wood; it is opake, elastic, of a slaty curve, and irregularly fibrous texture. *Anal.* Carbonate of magnesia 22. Silex 62. Argill 2,8. Carbonate of lime 10. Oxyde of iron 3,2. *Bergman.*

m. *Anctynolite*, amorphous, lamellated, striated, and fibrous; in compressed hexahedral prisms with smooth surfaces. *Anal.* Carbonate of magnesia 20. Silex 64. Carbonate of lime 9,3. Argill 2,7. Oxyde of iron 4. *Bergman.*

n. *Jade*, of a dark leek green or bluish colour; surface smooth, sometimes uneven; in detached round masses, or inhering in rocks; of a splintery texture; feels greasy. *Anal.* Carbonate of magnesia

magnesia 38. Silex 47. Carbonate of lime 2. Argill 4. Oxyde of iron 9. *Hoepfner.*

o. *Baikalite*, crystallized in tetrahedral prisms, entire or truncated, with oblique prisms; and in hexahedral prisms. *Anal.* Magnesia 30. Silex 44 Lime 20. Oxyde of iron 6. *Lowitz.*

p. *Boracite*, found only in the mountain of Kalkberg, near Luneburg; colour greyish white, inclining to purplish; crystallized in cubes, with truncated edges and angles; does not effervesce, or dissolve in acids, unless heat be applied. *Anal.* Boracic acid 68. Magnesia 13. Lime 11. Silex 1. Argill 1. Iron 1. *Westrumb.*

E. *ARGILL*, or Alumine, aluminous, or argillaceous earth; the earth contained in sulphate of alumine, or common alum, has no taste, and is insoluble, hardens in the fire, although divisible in water, contracting in it's dimensions, and when exposed to an extreme heat, becomes so hard as to strike fire with steel.

a. *Sulphate of Alumine*, or alum, is an earthy salt, which consists of the sulphuric acid and alumine.

b. *Carbonate of Argill*, Lac lunæ. Colour pure white, found in nests, in compacted, rounded, kidney-form masses. Consists of carbonic acid, argill and some lime. *Sckreiber.*

c. *Clay*, *Pipe Clay*, white, very little greasy to the

the touch, friable and ſtains the fingers ſlightly, diffuſible in water and ductile; effervefces a little with acids. *Anal.* Carbonate of argill and ſilex.

d. *Porcelain Clay,* white, and greyiſh, ſoft to the touch; when mixed with and moulded into veſſels, and expoſed to a ſtrong heat, aſſumes the appearance of a ſemi-tranſparent glaſs; ſtrikes fire with ſteel; is not acted upon by acids; and preſerves it's ſhape. *Anal.* Argill 60. Silex 20. *Wedgewood.*

e. *Lithomarga,* Mountain ſoap, friable, indurated, amorphous; texture frequently poliſhed, ſlaty, conchoidal; colour white, yellowiſh or reddiſh white; emits a phoſphoreſcent light, when rubbed in the dark with a pin; ſoft and ſoapy to the touch, abſorbs water rapidly, and retains it for a long time. *Anal.* Argill 11. Silex 60. Carbonate of lime 5,7. Carbonate of magneſia 0,5. Oxyde of iron 4,7. Air and water 18. *Bergman.*

f. *Fullers Earth,* greeniſh white colour, or greyiſh white, receives a poliſh from friction; does not adhere to the tongue; feels ſomewhat greaſy; falls into powder in water. *Anal.* Argill 0,25. Silex 0,51. Carbonate of lime 0,03. Carbonate of magneſia 0,007. Oxyde of iron 0,03. Moiſture and air 0,15. *Bergman.*

g. *Bole,* particles very ſmall, and cohere

ſlightly; ſoftens eaſily in water, but is not fit for pottery; feels greaſy; acquires a poliſh by light friction; of different colours, the red called *Armenian bole*, uſed for red pencils. The *Terra Lemnia*, which belongs to this tribe, contains ſilex, argill, magneſia, calcareous earth, iron and water. The indurated bole, or *red chalk*, is hard and brittle, and intimately united with oxyde of iron. *Anal.* Argill 19. Silex 47. Carbonate of lime 5,4. Carbonate of magneſia 6. Oxyde of iron 5,4. Water and air 17. *Bergman.*

h. *Tripoli*, ſo called from the place whence it was firſt brought to us; colour whitiſh, grey or ochre yellow; is found ſolid; has a dull and earthy appearance when broken; is ſoft and ſandy between the teeth; abſorbs water with a noiſe, during which air bubbles are expelled; does not efferveſce with acids, unleſs mixed with marl, which it often is. *Anal.* Argill 7. Silex 90. Iron 3. *Hasse.*

i. *Lepidolite*, amorphous, foliated. *Anal.* Argill 38,25. Silex 54,5. Oxyde of iron and manganeſe 0,075. Water and air 2,5. *Klapr.* Silex 54. Alumine 20. Fluate of lime 4. Oxyde of manganeſe 3. Oxyde of iron 1. Potaſh 18. *Vauquelin. Lelievre.*

k. *Sappare*, Cyanite, compoſed of long laminæ; white, with ſhades of ſky or Pruſſian blue; luſtre like mother of pearl; it is rather brittle, ſemi-tranſparent, ſoft, and can be ſcraped with a knife; feels to the touch like talc. *Anal.* Argill 67.

Silex

Silex 13. Magnefia 13. Iron 5. Lime 2. *Saussure, jun.*

l. *Mica*, lamellated ftructure, the laminæ fometimes of confiderable fize, femi-tranfparent, and the *Muscovy glass* in thefe laminæ is quite tranfparent, and is ufed for windows and lanterns; flexible and elaftic; is found of various colours, from filver white to blackifh. *Anal.* Argill 28. Silex 38. Magnefia 20. Oxyde of iron 0. *Kirwan.*

m. *Micarelle*, found in granite; of a brownifh black colour; texture foliated. *Anal.* Argill 63. Silex 29. Oxyde of iron 7. *Klapr. Anal.* Argill 63. Silex 29. Oxyde of iron 7. *Kirwan.*

n. *Hornblende*, Bafaltic hornblende, Common hornblende, Schiftofe hornblende, found of various degrees of hardnefs, but never fo hard as to ftrike fire with fteel; blackifh or deep green colour; cryftallized in fix-fided prifms, the ends acuminated by three faces; and alfo terminating varioufly; cryftals internally fhining; the texture exhibits diverging radiations. *Anal.* Argill 27. Silex 58. Lime 4. Magnefia 1. Lime 9. *Bergman.* Conf. *Kirwan.*

o. *Refplendent Hornblende*, Labrador hornblende, Schiller fpar, colour greyifh black, fometimes with a fhade of copper red, refembling Labrador ftone; texture lamellated; the lamellæ curved. *Anal.* Argill 17. Silex 42. Magnefia 11. Iron 28. *Gmelin.*

p. *Basalt*,

p. *Basalt*, Trap, Wakken, Mullen, Krag; colour blackiſh green, and ſometimes deep black; cryſtallized in ſix and eight ſided priſms, cuneated on the ends; it's ſurface ſhining on being broken; when longitudinally, the lamellæ are parallel. *Anal.* Argill 15. Silex 50. Carbonate of lime 8. Iron 28. Magneſia 2. *Bergman.* Conf. *Withering.*

q. *Calp*, ſlaty, amorphous. *Anal.* Argill, ſilex, and iron, with fifty per hundred weight of carbonate of lime. *Kirwan.*

r. *Argillaceous Schistus*, Argillite, Killas, Grapholite; colour bluiſh or bluiſh grey, purpliſh or black; amorphous and ſlaty. *Anal.* Argill 25. Silex 60. Magneſia 9. Iron 6. And ſome petroleum. *Kirwan.*

ſ. *Novaculite*, Turkey hone, amorphous and ſlaty.

F. *SILEX*, is rough to the touch, cuts glaſs, and ſcratches metals; is infuſible, and inſoluble in moſt acids, except the fluoric, forming Derbyſhire ſpar; ſoluble in alkalies, with the aſſiſtance of heat forming glaſs; found in moſt ſtones that ſtrike fire with ſteel.

a. *Diamond*, cryſtallized in octohedrons* and their varieties; and in dodecahedrons† and icoſa-

* Octohedrons, ex οκτω, eight; et εδρα, baſis.

† Dodecahedral, ex δωδεκα, twelve; et εδρα, baſis.

hedrons

hedrons* and their varieties, indeterminate; found in a ſandy earth in the hither peninſula of India, in the iſland of Borneo, and in the Brazils. Diamonds are of a lamellated texture, and may be ſplit by a blow in a proper direction; cut all other ſubſtances, and take a moſt exquiſite poliſh.

b. *Sapphire* is next in hardneſs to the diamond; the ſtone worn by the high prieſt of the Jews; of a ſky blue colour, ſometimes inclining to pink, ſometimes almoſt white, and then called *lux sapphire*; it is found cryſtallized in lengthened hexahedral priſms joined baſe to baſe. *Anal.* Silex 35. Argill 58. Carbonate of lime 5. Iron 2. *Bergman.*

c. *Topaz of Brazil*, colour various, the true oriental is almoſt colourleſs, or pale yellow; the Brazil is of a fine, yellow, tranſparent amber, cryſtallized in tetrahedral rhomboidal priſms, terminated by tetrahedral pyramids.

d. *Topaz of Saxony*, cryſtallized in four-ſided rhomboidal priſms, terminating from the ſides in flat four-ſided pyramids, longitudinally ſtriated, and of a foliaceous texture. *Anal.* Silex 52. Argill 44. Lime 2. Iron 0,03. *Wiegleb.*

e. *Beryl of Siberia*, colour greyiſh green, verging on the apple green, more rarely bluiſh green; cryſtallized in equi-angular hexahedral priſms longitudinally ſtreaked: becomes electric by

* Icoſahedrons, ex εικοσι, viginti; et εδρα, baſis.

friction,

friction, but one of it's poles is attractive whilst the other is repulsive. *Anal.* Silex 61. Argill 29. Calx 2. Iron $\frac{1}{2}$. Seven parts lost. *Herman.*

f. *Ruby*, of a crimson, deep red colour, which it retains in the fire; crystallized in octohedrons and their varieties, and double four-sided, seldom six-sided pyramids. *Anal.* Silex 16. Argill 76. Lime 1. Iron 3. *Klaproth.*

g. *Emerald*, Smaragdite, of a pure green colour, which it loses in the fire, as well as it's weight; crystallized in hexagonal prisms, either perfect or truncated on the angles or edges, terminating in truncated pyramids; the crystals mostly smooth on the surface, shining and transparent in various degrees. The light green coloured oriental emerald is reckoned the best. *Anal.* Silex 24. Argill 60. Lime 8. Iron 6. *Klaproth.* The Smaragdite has been analysed by *Vauquelin.* The *Green and White Smaragdite* of Corsica. Silex 50. Alumine 11. Magnesia 6. Lime 13. Oxyde of iron 5,5. Oxyde of copper 1,1. Oxyde of Chrome 4. The *green* Smaragdite. Silex 51. Alumin 13,5. Magnesia 5. Lime 14,5. Oxyde of iron 8. Oxyde of copper 0,5. Oxyde of Chrome 4. The *Grey.* Silex 50. Alumine 7. Magnesia 8. Lime 17. Oxyde of iron 14,5.

h. *Aqua Marine*, Aigue marine, crystallized in large, perfect, six-sided prisms; colour of a pale green *Anal.* Silex 64. Argill 24. Lime 8. Iron 1,5. *Bindheim.*

i. *Crysolite*,

i. *Crysolite*, colour yellowish green, inclining to yellowish brown; crystallized in hexahedral prisms with corresponding pyramids. *Anal.* Silex 15. Argill 64. Lime 17. Iron 1. *Achard.*

k. *Hyacinth*, of a peculiar yellowish red; in general retain their colour in fire; crystallized in dodecahedrons with unequal rhombic faces. *Anal.* Silex 25. Argill 40. Carbonate of lime 20. Iron 13. *Bergman.*

l. *Hyacinth of Vesuvius*, is generally of a deeper colour than the oriental, and is less hard and more fusible. Crystallized in tetrahedral prisms, truncated at their angles, terminated by tetrahedral pyramids, truncated at their summits.

m. *Olivin*, found imbodied in basaltes; of an olive green colour; crystallized in six, seldom in four, sided prisms; mostly rectangular, with or without pyramids. The crystals are longitudinally striated; have a vitreous appearance. *Anal.* Silex 54. Argill 40. Iron 4. *Gmelin.*

n. *Garnet*, Carbuncle, of a red colour, or some of it's shades, a greenish white, and of various degrees of transparency; does not lose it's colour in the fire, but becomes heavier, and is strongly attracted by the magnet. Crystallized and presenting either 12 rhomboidal planes, or 24 trapezoidal, or 36 planes, of which 12 are rhombs, and the 24 others elongated hexagons, interposed between these rhombs, and some other variations. *Anal.* Silex 48. Argill 30. Lime 11. Iron 10. *Achard.*

o. *White*

o. *White Garnet,* Veſuvian, always exhibits 24 trapezoidal faces, found in volcanic productions; is not fuſible by fire like the red. *Anal.* Silex 55. Argill 89. Lime 6. *Bergman.*

p. *Tourmalin,* of a deep brown, hyacinth, or green colour; does not ſtrike fire with ſteel; can be ſcratched by a knife; cryſtallized in ſix or nine-ſided priſms, longitudinally ſtreaked, and terminating in three-ſided pyramids with pentagonal faces. Is electric when heated. *Anal* Silex 37. Argill 29. Lime 15. Iron 9. *Bergman.*

q. *Schorl,* of various colours, red, black, greyiſh white, but generally green; ſomewhat tranſparent; ſcratches glaſs, and moſt varieties ſtrike fire with ſteel; cryſtallized in ſix or nine-ſided priſms, with three-ſided pyramids. *Anal.* Silex 52. Argill 37. Lime 5. Magneſia 3. Iron 3. *Chaptal.*

r. *Thumerstein,* Violet ſchorl, vitreous appearance, more or leſs tranſparent; cuts glaſs, and ſtrikes fire with ſteel; found in flat rhomboidal cryſtals, with the two oppoſite edges a little truncated; the ſurface of the ſides ſtreaked, and the ſurface of truncation ſmooth and ſhining. *Anal.* Silex 52. Argill 25. Lime 9. Iron 9. And ſome manganeſe. *Klaproth.*

ſ. *Schorlite,* cryſtallized, indeterminate. *Anal.* Silex 50. Argill 50. *Klaproth.*

t. *Rubellite,* cryſtallized in diverging ſtriated priſms,

prisms, with trihedral summits. *Anal.* Silex 57. Argill 35. Oxyde of iron and manganese 5. *Bindheim.*

u. *Amethyst*, crystallized in hexahedral prisms, with corresponding pyramids; of various degrees of transparency, and colour. *Anal.* Silex 30. Argill 60. Lime 22. Iron 1,66. *Achard.*

v. *Quartz*, crystallized in double hexahedral pyramids, with or without an intermediate prism; of various shapes; strikes fire with steel; is soluble in fluoric, but not in nitric or muriatic acid; *Anal.* Silex 93. Argill 6. Lime 1. *Bergman.*

w. *Prase*, of a leek green colour; crystallized in six-sided prisms, acuminated by six faces, proceeding from the sides of the prisms.

x. *Elastic Quartz*, amorphous, angular. *Anal.* Silex 0,965. Argill 0,025. Iron 0,01. *Klaproth*,

y. *Obsidian*, Iceland Agate, of a greyish or blackish colour; found in irregular shaped pieces. *Anal.* Silex 69. Argill 22. Iron 0,009. *Bergman.*

z. *Calcedony*, Cornelian, Agate, called Calcedon, from the name of a place; usually found in solid masses, of a globular, kidney-like, or stalactical figure, or in the state of pebbles; sometimes filiform, tabular or cellular; of a greyish, yellowish, white, or milk white colour. *Anal.* Silex 84. Argill 16. *Bergman.*

a a. *Chrysoprase*, of a green, or greenifh white colour; found in folid maffes, and breaks with an even furface, having a dull appearance; does not ftrike fire with fteel. *Anal.* Silex 0,96. Oxyde of Nickel 0,01. Lime 0,0083. Argill 0,0083. Oxyde of iron 0,0083. *Klaproth.*

b b. *Hyalite*, amorphous, lamellar, compact, colour pure white, fracture even, lamellated, fometimes conchoidal. *Anal.* 57. Argill 18. Lime 15. And a very little iron. *Link.*

c c. *Opal*, Semi-opal, hard and femi-tranfparent; reflects the light in various hues; found of a milk white, bluifh white and greenifh colour, rarely purple; takes a fine polifh, but is not fo hard as to ftrike fire with fteel. *Anal.* Silex 98,75. Argill 0,01. Oxyde of iron 0,01. *Klaproth.* Of this clafs are *Hydrophanes*, which become tranfparent by placing them in water.

d d. *Pitchstone*, cryftallized in hexahedral prifms with trihedral pyramids; it's various colours are white, yellow, brown, greenifh, feldom tranfparent; fcratches glafs. *Anal.* Silex 73. Argill 18. Iron 00,58. *Wiegleb.*

e e. *Cats-eye*, amorphous, compact, colour greyifh white, yellowifh or reddifh brown, or ftriped with thefe colours, particularly when polifhed. Found in blunt or rounded fragments, and comes from Ceylon.

f f. *Flint*, ftrikes fire with fteel, generally of a yellowifh yellowiſh ſmoke grey, or black grey colour; cryſtallized in double trihedral pyramids; in maſſes of various ſizes, in ſtrata of chalk; when broken, exhibits a conchoidal ſurface, and the fragments are ſharp edged, but of an indeterminate ſhape; ſemi-tranſparent. *Anal.* Silex 60. Argill 18. Lime 2. *Wiegleb.*

g g. *Hornstone, Siliceous Schistus,* or *Basanite, Hornslate. Petrosilex.* Cryſtallized in hexahedral priſms, with or without pyramids, in double trihedral pyramids, in cubes; amorphous, compact; ſlaty; colour various, yellowiſh white, bluiſh, pearl, green inclining to red, browniſh, roſe colour, blackiſh brown. *Anal. Hornstone.* Silex 72. Argill 22, Carbonate of lime 6. *Kirwan. Basanite.* Silex 73. Lime 10. Magneſia 0,046. Iron 3. Coal 5. *Weigleb. Hornslate.* Silex 73. Argill 24. Iron 3. *Wiegleb.*

h h. *Jasper, Egyptian Pebble,* found in irregular hexahedral priſms; amorphous, in large maſſes and nodules, and of all colours. *Anal.* Silex 54. Argill 30. Iron 16. *Delametherie.*

i i. *Porcelanite,* Porcelain Jaſper, of a pearl grey, or lavender blue colour; found in compact layers. It's ſurface, when newly broken, is a little ſkinny and gibbous.

k k. *Heliotropium, Bloodstone,* of a leek green colour, with blood-red ſpots or veins; is ſemitranſparent; exhibits a conchoidal ſurface when broken.

It may be obſerved that Agates do not form a diſtinct ſpecies of ſtone, but conſiſt of quartz, cryſtal, hornſtone, flint, calcedony, amethyſt, jaſper, cornelian, heliotropium, jade, &c. in various combinations, and may be exhibited in the following manner, from *Kirwan*.

The clear pellucid white denotes - -	cryſtal or quartz.
The yellowiſh white -	amethyſt, calcedony;
The greyiſh white -	quartz;
The milk white -	jaſper, amethyſt;
The yellowiſh grey -	quartz, flint;
The ſmoke grey -	calcedony, flint;
The pearl grey -	calcedony;
The greeniſh grey -	calcedony, jaſper;
The bluiſh grey -	calcedony.
The honey yellow -	flint, jaſper;
The yellow - -	jaſper;
The ochre yellow -	calcedony;
The orange yellow -	calcedony;
The yellowiſh brown	cornelian;
The reddiſh brown -	hornſtone, cornelian;
The fleſh red - -	cornelian;
The blood red -	jaſper, cornelian;
The brick red -	jaſper.
The violet blue -	amethyſt;
The browniſh green -	heliotropium;
The leek green, and feeling ſomewhat greaſy -	jade.

11. *Feltspar*, *Adularia*, or Moonſtone, cryſtallized in tetrahedral priſms, truncated obliquely, and

and their varieties; tranſparent; ſhining lamellæ, placed one over the other, often in an irregular manner; but it s figure approaches always either to the cubical or rhomboidal. *Anal.* Silex 62. Argill 17. Lime 6,5. Baroſelenite 2. Magneſia 6. Iron 1,4. *Westrumb.*

m m. *Common Feltspar*, of a pale fleſh colour, yellowiſh, grey, milk white; rarely of a vivid green or bluiſh colour; opaque, compact, ſtrikes fire with ſteel. *Anal.* Silex, argill, with lime and magneſia, or baryte. *Kirwan.* Silex 62,83. Alumine 17,002. Lime 3. Oxyde of iron 1. Loſs. 16,015. *Vauquelin.*

n n. *Labrador Feltspar*, amorphous, foliated, of various hues, light or deep grey, but when held in certain directions to the light, reflects lazuli-blue, green, and lemon yellow, coppery or violet colour, the ſurface ſhining, texture foliated, fragments rhomboidal. *Anal.* Silex 69. Argill 13. Sulphate of lime 12. Oxyde of copper 0,7. Oxyde of iron 0,04. *Bindheim.*

o o. *Argentine Feltspar*, cryſtallized, as common feltſpar; amorphous, foliated. *Anal.* Silex 46. Argill 36. Oxyde of iron 16. *Dodun.*

p p. *Staurolite*, cryſtallized in tetrahedral priſms, with tetrahedral pyramids, either ſingle, or croſſing each other at right angles. *Anal.* Silex 44. Argill 20. Baryte 20. Water 16. *Westrumb.*

q q. *Lapis*

q q. *Lapis Lazuli*, is opaque, compact, and never found cryſtallized; amorphous, ſeldom pure; of a ſky-blue colour, which it retains in the fire for a long time, but at length becomes brown. *Anal.* Silex, lime, ſulphate of lime, and iron. *Margraff.*

r r. *Prehnite*, ſo called by Werner, after Capt. Prehn, who brought it firſt to Europe from the Cape in 1783. Cryſtallized in tetrahedral priſms; indeterminate; amorphous, foliated; ſemi-tranſparent and brittle; ſurface ſhining, with a lamellated, or fibrous texture; of a greeniſh grey colour. *Anal.* Silex 44. Argill 30. Lime 18. Iron 5. Water and air 2. *Klaproth.*

ſ ſ. *Adelite*, of particular ſhapes, tuberous. *Anal.* Silex 62 to 69. Argill 18 to 20. Lime 8 to 16. Water 3 to 4. *Bergman.*

t t. *Zeolite*, opaque, ſeldom ſemi-tranſparent; hard, but rarely ſo hard as to ſtrike fire with ſteel; of a pale green, ſilver white and honey colour, ſometimes of a bluiſh or coppery hue; cryſtallized in cubes; it is found compact, fibrous, radiated, lamellated, uniform, ſtalactical, in drops, and of a capillary ſhape, in which it is very beautiful; in ſix-ſided and flat priſms. *Anal.* Silex 50. Argill 20. Lime 8. Water 22. *Pelletier.*

u u. *Siliceous Spar*, cryſtallized in tetrahedral, and in hexahedral priſms. *Anal.* Silex 61,1. Lime 21,7. Argill 6,6. Magneſia 5. Oxyde of iron 1,3. Water 33. *Bindheim.*

v v. *Rose*

v v. *Rose Spar*, Red Stone, cryſtallized, indeterminate; amorphous, thick foliated.

C. *ADAMANTINE EARTH*, or CORUNDUM, diſcovered by Klaproth, in adamantine or diamond ſpar.

a. *Adamantine Spar*, of a greyiſh colour, inclining to greeniſh white, chocolate brown, &c. cuts glaſs, and ſcratches other gems; ſtrikes fire with ſteel; is not affected by ſulphuric, nitric, or marine acid; exhibits ſix-ſided ſhort rhomboidal priſms, rounded on the top, of a lamellated texture, ſhining in certain directions. *Anal.* Adamantine 68. Silex 31,5. Iron and nickel 00,05, *Klaproth.*

	Corundum of China.		of Peninſula of India.
Argillaceous earth	89,50	—	84,0
Siliceous earth	5,50	—	6,50
Oxyde of iron	1,25	—	7,50
Loſs —	3,75	—	2,0
	100,0		100,0

Chs. Greville.

H. *JARGON EARTH*, diſcovered by Klaproth, in the ſtone called jargon, or zirgon, of Ceylon.

a. *Jargon*, cryſtallized in ſhort tetrahedral priſms, with tetrahedral pyramids; indeterminate, in ſmall grains, and ſmall flat pebbles; ſcratches glaſs;

glaſs: bears a greater heat than the diamond before it is conſumed. *Anal.* Jargon earth 68. Silex 31. Iron and nickel 5. *Klaproth.*

I. *SIDNEIAN EARTH*; diſcovered by Wedgewood; it is contained in a compound ſubſtance from Sidney Cove in South Wales; conſiſting of fine white ſand, ſome colourleſs mica, a few black particles reſembling black lead, and a white argillaceous earth, and from which the new earth is extracted by the marine acid.

a. *Sidneia*, amorphous, looſe. *Wedgewood.*

II. Mixed.

A. CALCAREOUS. Calcareous earth, in it's moſt common ſtate, is called limeſtone, or ſpar; it is in this ſtate that it is combined with a peculiar acid, expellable by heat, and hence called aërial acid, fixed air, or carbonic acid; it effervesces with acids in general: it's more mixed ſtate is now conſidered.

a. *Marl*, Argillo-Calcite, ſemi indurated or earthy form, of a yellowiſh grey, or yellowiſh white colour; looſe texture, ſtrongly effervesces with acids; in water ſoon falls into powder. In a ſtony or indurated ſtate, fracture earthy, ſometimes ſplintery, or conchoidal, frequently ſlaty; colour various, yellow, grey, brown or bluiſh. *Anal.* Carbonate of lime 60 to 80. Remainder argill; and ſilex. *Kirwan.* In it's ſtony or indurated

durated ſtate, colour yellow, grey, brown, or bluiſh; texture ſplintery or conchoidal, frequently ſlaty; penetrated by bitumen, ſulphur, and pyrites. *Kirwan.*

b. *Limestone with argyllite.*

c. *Siliceous Limestone.*

d. *Ferruginous Limestone.*

e. *Gypsum with Calcareous Spar*; Gypſum is frequently penetrated with this ſpar, ſwineſtone, ſandſtone, and perhaps with ſtrontian; which cauſe a change in it's hardneſs and gravity, and of colour from blackiſh to whitiſh grey.

f. *Gypsum with Marl;* ſometimes diſguiſed with iron; colour reddiſh brown, and reddiſh black; effervefces with acids; gives a red ſtreak.

B. MAGNESIAN.

a. *Calciferous Asbestinite.*

b. *Steatite with Argill.*

c. *Serpentine with Hornblende.*

d. *Siliciferous Potstone;* bluiſh black, mixed with white; it's fracture uneven, partly ſplintery, partly ſlaty; the quartz in many parts viſible in the veins.

e. *Ferruginous Steatite.*

C. ARGILLACEOUS.

a. *Calciferous Argillite;* colour dark, blackiſh, or bluiſh grey, or grey with black or deep blue blotches, or reddiſh or yellowiſh grey; more rarely greeniſh grey; fracture ſlaty, but of ſingle laminæ, ſplintery, or conchoidal.

b. *Talcose Argillite;* colour whitiſh, or bluiſh, or greeniſh grey; often inveſted with foliated ſteatites; found in large maſſes; fracture ſlaty, curved.

c. *Siliciferous Argillite;* penetrated with ſiliceous ſchiſtus, ſand, jaſper, baſanite, or quartz; colour dark, blackiſh, or bluiſh grey, rarely greeniſh grey; fracture ſlaty, but of ſingle laminæ; ſplintery, or conchoidal.

d. *Ferruginous Argillite*; penetrated with calces of iron, of a bluiſh brown colour; externally, it has ſome luſtre; internally, none; ſomewhat of a ſtriated, as well as of a ſlaty appearance.

e. *Hornblende with Garnet;* of a dark greeniſh red, or reddiſh dark green, according to the proportion of garnet; fracture earthy, or fine ſplintery.

f. *Hornblende Slate with Talc or Mica*; colour greeniſh grey; fracture ſlaty, and thin, the laminæ not eaſily ſeparable; found in great plenty

at

at Holyhead, and has for the moſt part thick layers of quartz intercepted between it's laminæ.

g. *Hornblende Slate with Quartz;* colour ochre yellow, with black ſhining ſtreaks of hornblende; fracture ſchiſtoſe; feels ſandy; reſembles gneiſs; the yellow colour ſeems to proceed from a ferruginous quartz.

h. *Trap with Hornblende;* colour bluiſh, or greeniſh black; fracture fine ſplintery; contains ſome grains of magnetic ironſtone.

i. *Trap with Mullen;* colour greyiſh black, with a ſhade of red, fracture uneven and fine ſplintery.

k. *Trap with Krag;* colour greyiſh black, with a ſhade of red; fracture uneven and fine ſplintery.

l. *Siliciferous Trap;* colour blackiſh grey, or dark iron grey, with numerous rounded white ſpecks, as ſmall as the point of a pin; it is alſo full of rifts, and theſe exhibit a bluiſh or reddiſh ferruginous illinition; fracture uneven and ſplintery.

m. *Ferruginous Clays;* ſome are ſo penetrated with calces of iron, as ſcarcely to indicate a ſlaty texture; colour bluiſh brown; externally ſome luſtre; internally none; readily imbibes water.

n. *Mullen with Asbestinite;* colour of a reddiſh grey when firſt broken, with very little luſtre; fracture uneven and earthy.

D. *Sili-*

D. SILICEOUS.

a. *Earthy Quartz;* ſtructure looſe as that of moſs, which it reſembles; exceedingly brittle, and light as pumice; colour white, greyiſh, and dark.

b. *Ferruginous. Quartz;* hard, fracture ſplintery; colour brown, or browniſh purple; ſometimes ſchiſtoſe, ſometimes of the texture of moſs.

c. *Earthy Quartz with Actinolite;* greeniſh grey colour, with yellowiſh green lumps, and ſpecks and blotches of red; fracture ſplintery and uneven.

d. *Earthy Hornstone*, reſembles earthy quartz, but of a finer and cloſer grain; colour reddiſh grey mixed with green, browniſh or blackiſh; ſtructure thick ſlaty; fracture ſplintery.

e. *Ferruginous Hornstone;* hard and cloſe grained, colour reddiſh brown; gives a pale red ſtreak.

f. *Siliceous Schistus with Limestone;* of a dark yellowiſh grey colour; fracture ſplintery; the grain very fine and cloſe.

g. *Siliceous Schistus with Argillite;* colour light bluiſh grey or purpliſh grey; fracture fine ſplintery; in the groſs, ſlaty.

h. *Siliceous Schistus with Mullen;* colour bluiſh grey, it's ſurface often ſtained reddiſh or yellowiſh brown; fracture fine, ſplintery, and uneven.

i. *Pitchstone with Opal;* of a pearl grey, reſembling porcelanite.; penetrated with lumps of opal; brittle; imbibes water ſlowly; decrepitates ſtrongly by heat, and then whitens, and remains in a looſe powder.

III. Aggregated.

A. CALCAREOUS.

a. *Calcareous Sandstone*; cryſtallized in rhomboids; amorphous; colour generally white or grey, or yellowiſh white; ſurface rough; fracture earthy or ſlaty; often ſpangled with mica. *Anal.* Carbonate of lime 37,5. Silex 62,5. *Lassone.*

b. *Calcareous Breccia*; conſiſting of fragments of marble in a calcareous cement.

B. MAGNESIAN.

a. *Potstone Porphyry*; amorphous, undulatingly foliated; greeniſh, reddiſh, or yellowiſh grey; or ſpeckled earth; red or leek green; containing potſtone and feltſpar.

b. *Serpentine Porphyry*; amorphous, compact; dark or olive green, grey, or reddiſh; fracture ſplintery; found near Florence, containing ſerpentine and feltſpar.

C. ARGIL-

C. ARGILLACEOUS.

a. *Argillaceous Sandstone*; amorphous, flaty, and compact; does not effervesce with acids; often shot through with mica; make mill-stones, filtering-stones, and coarse whetstones; composed of argillaceous cement, with fragments of quartz, feltspar, and flint.

b. *Ruddle Stone*; amorphous, flaty, and compact; containing argillaceous cement, with quartz, siliceous schistus, or hornstone and argillite.

c. *Argillaceous Porphyry*; amorphous; contains feltspar in indurated clay; hornblende, trap, wakken, mullen, krag, or argillite.

d. *Amigdaloid*; amorphous; containing rounded masses of calcedony, agate, zeolite, calcareous spar, lithormaga, steatite, green earth, &c. in an argillaceous basis.

e. *Schistose Mica*; amorphous, flaty; contains mica and quartz.

D. SILICEOUS.

a. *Granite*; amorphous, compact, flaty; chiefly composed of quartz, feltspar, and mica; the compact parts generally irregularly mixed, and of various sizes.

b. *Sienite*; amorphous, compact, flaty; composed of quartz, feltspar, and hornblende; or of quartz,

quartz, feltſpar, hornblende, and mica; ſometimes found ſlaty.

c. *Granatine;* compoſed of quartz, feltſpar, ſchorl; quartz, feltſpar, garnet; &c. &c. &c. *Saussure.*

d. *Granitell;* or Grunſten; amorphous; compoſed of quartz, feltſpar; quartz, ſchorl; quartz, mica, &c. *Kirwan.*

e. *Granalite*; amorphous; compoſed of quartz, feltſpar, mica, ſchorl; quartz, feltſpar, mica, ſteatite, &c.

f. *Gneiss;* amorphous; ſlaty, fibrous, lamellated; contains quartz, feltſpar, and mica. *Werner. Saussure.*

g. *Siliceous Porphyry;* amorphous; compact, ſlaty; conſiſting of cryſtals of feltſpar in a baſis of jaſper, hornſtone, pitchſtone, obſidian, ſiliceous ſchiſtus, ſchiſtoſe hornblende or feltſpar.

h. *Pudding Stone*; amorphous; conſiſting of rounded pebbles in a ſiliceous cement.

i. *Siliceous Sandstone;* amorphous; the grains often ſlightly compacted together, and eaſily breaking into ſand; of this kind is *Salindre*, conſiſting of calcareous grains, inſerted in a ſiliceous cement.

k. *Siliceous Breccia;* amorphous; conſiſting of angular fragments of ſiliceous ſtones in a ſiliceous cement.

CLASS

CLASS III.

METALS.

THE most ponderous of all mineral bodies, fusible, but resuming their original properties when cold, even after calcination, by the addition of inflammable matter, or oxygen gas: in their purest metallic state, they possess neither taste nor smell. There are 23 metals with which we are acquainted, of these 8 are considered as ductile, or entire metals; and 15 as fragile or semi-metals; of the former are, 1. Platina. 2. Gold. 3. Quicksilver or Mercury. 4. Silver. 5. Lead. 6. Copper. 7. Iron. 8. Tin. Of the fragile are, 1. Bismuth. 2. Nickel. 3. Arsenic. 4. Cobalt. 5. Zinc. 6. Antimony. 7. Manganese. 8. Scheele, Wolfram, or Tungsten. 9. Uranite. 10. Molybdena. 11. Menachanite. 12. Sylvanite. 13. Titanite. 14. Chrome. 15. Tellurium.

From recent experiments made on the Carbonates of Barytes, Magnesia, and Chalk, three new metals are supposed to exist, which Mr. Tondi, the discoverer, has named *Barbonum*, *Austrum*, and *Parthenum*; for which Mr. Kerr would substitute the names of *Barytum*, *Magnesium*, and *Calcum*,

Calcum, as more consistent with the new chemical nomenclature; but further experiments are requisite to determine the truth of those already made, as they are doubted by some eminent chemists.

I. Ductile.

A. PLATINA, Plata Silver, or Platina de Pinto; found only in Spanish America in small grains among the gold mines there; introduced into England in the year 1749 It is usually united with iron, and in it's crude state is 18 times heavier than water, and when pure is heavier than gold, it's specific gravity being $21\frac{1}{2}$. With a strong white heat it's parts will adhere together by hammering. The best solvent is aqua regia, composed of equal parts of the nitrous and marine acids.

B. GOLD is a yellow metal, 19 times heavier than water; unalterable by air, or fire; soluble in aqua regia, composed of three parts of nitrous and one of marine acid, and may be precipitated by various substances; lime and magnesia precipitate it in the form of a yellow powder; the nut-gall precipitates it of a reddish colour; the volatile alkali of a brown, yellow, or orange colour, known by the name of fulminating gold, and this powder weighs a fourth more than the gold made use of; liver of sulphur precipitates it, by the alkali uniting with the acid, when it falls down combined with the sulphur; lead, iron, and silver precipitate it of a deep and dull purple colour; copper and iron throw it down in it's metallic state. A plate of tin immersed in a solution of gold, affords a purple powder, called the purple powder of Cassius, which is used to paint in enamel. Most metals unite with gold in fusion, hence it is generally found alloyed with silver, copper, iron, or other metals.

a. *Native Gold*; found in it's metallic state, crystallized with silver, copper, or iron.

b. *Grey Ore*; combined with sulphur, antimony, arsenic, lead, iron, and silver. *Born.*

c. *White Ore*; combined with silver, bismuth and sulphur.

The largest quantities of gold are brought from the Brazils and the Spanish West Indies. This metal is found also in Hungary, Transylvania, and in many other parts of Europe, in red, yellow, black, or iron-coloured sands; it is met with likewise in some rivers, as the Tagus, Ganges, Rhine, Saale, Niger, Danube, &c. called river, wash-gold, or gold-dust.

C. QUICKSILVER.

QUICKSILVER, or Mercury; the most fluid of all metals, not taking a solid state until cooled

to

to the 29th degree below o on Fahrenheit's thermometer, when it becomes malleable; it is 14 times heavier than water, easily divisible, and evaporates in a heat below ignition; rapidly soluble in the nitrous acid. Mercury is found in Hungary, Transylvania, Carinthia, Bohemia, France, Spain, Sweden, Peru, and probably in the East Indies and Japan.

a. *Native Quicksilver*; fluid, or interspersed; sometimes called virgin mercury.

b. *Amalgamized Mercury*; alloyed with silver; found on Moschellandsberg and Stahlberg in Deuxponts, near Sahlberg in Sweden, near Zlana in Hungary, chiefly on a grey indurated clay.

c. *Cinnabar*; combined with sulphur, in various forms, scaly, granular, indeterminate.

d. *Hepatic Ore*; Mercury combined with sulphuret of potash or soda.

e. *Cupreous Mercury*; Quicksilver combined with copper, chiefly on lapis ollaris and quartz. Found in the mines near Moschellandsberg.

D. SILVER.

SILVER, when pure, is 11 times heavier than water; the whitest of all metals; harder, but less malleable than gold; soluble in the nitrous and vitriolic acids. Silver is found in all countries, but most plentifully in Peru and Potosi.

a. *Native Silver*; found generally of about 16 carats ſtandard.

b. *Arsenical Silver*; cryſtallized with arſenic and iron.

c. *Horn Silver*; reſembling reſin in colour, containing muriatic acid, ſulphuric acid, iron, argill, and lime.

d. *Vitreous Silver*, or Glaſs ſilver ore, of a dark colour like lead ore, contains ſulphur, is ductile.

e. *Brittle Vitreous Ore*, contains iron, antimony, ſulphur, copper and arſenic.

f. *Red Silver Ore*, a brittle red-coloured ore, containing regulus of antimony and ſulphur, and ſulphuric acid. The ſilver is in proportion of 62 to 100. *Klaproth.*

g. *White Silver Ore*; combined with lead, antimony, ſulphur, iron, ſilex, argill; of a light grey colour, and of a dull ſteel-grained texture.

E. LEAD.

LEAD is above 11 times heavier than water, unſonorous, malleable, and very fuſible, ſoluble in all acids and alkaline ſolutions. It yields an oxyde more vitrifiable than the oxyde of any other metal, and a glaſs of a topaz-yellow colour.

a. *Native*

a. *Native Leaa*; found in Poland, Silesia, near Karthen, and in Monmouthshire; also in Carinthia, in Vivarais.

(a.) 2. *Native Oxyde of Lead*; crystallized in various forms; compact, or pulverulent. *Anal.* Lead 36. Oxygen 37. Iron 24. Argill 2. *Macquart.*

(a.) 3. *Molybdate of Lead*; crystallized in rectangular tables and their modifications.

b. *Lead Amalgam*; or lead combined with mercury.

c. *Red Lead Ore*, or Ferruginous oxyde of lead; resembling arsenic or realgar; crystallized, shining, striated on the surface; semi-transparent.

d. *White Opaque Lead Ore*, or Native White Lead; of a yellowish or greyish colour. Found at Blegberg.

e. *White Carbonate of Lead*, { Chalk of lead. Spathose lead. Mephite of lead.

Spathose lead ore, White transparent lead ore, or White lead spar; colour white or colourless, usually mixed with calcareous earth or clay. Found near Freiburg, Marienberg, in Siberia, Bohemia, Hungary, Tipperary in Ireland, and lead hills in Scotland. *Anal.* Lead 80. Carbonic acid 16, with some lime and argill. *Westrumb.*

f. *Black*

f. *Black Lead Spar*; of a yellowiſh green, or verdegris colour; of a truncated hexahedron form; found on the lead-hills.

g. *Brown Lead Ore*; of a reddiſh brown colour, in four-ſided priſms; found in Tipperary, alſo at Zſchopau.

h. *Native Glass of Lead*; depoſited upon calcareous ſpar, reſembling mica; found at Bergmanſtorff, near Zellerfeld, on the Harz.

i. *Antimonial Lead Ore*; of an iron grey colour. Found at Sahlberg in Sweden, in Siberia; at Lautenthal on the Harz; and in Hungary. *Anal.* Lead 40,50. Antimony 8,16. With ſome ſilver.

k. *Galena*; Sulphuret of Lead, or Sulphurated lead ore; one of the moſt common lead ores, combined with ſulphur, ſometimes ſilver, iron and antimony; it's matrix is generally heavy ſpar, fluor, quartz, coal, ſpathoſe iron ore, ſchiſtus and gneiſs. Found at Bleyſtedt in Bohemia, Frieberg in Saxony; Siberia, Sahlberg in Sweden, and in Derbyſhire. *Anal.* Lead 77. Sulphur 20. Silver 1. *Kirwan.*

l. *Pyritical Lead Ore*; combined with martial pyrites.

m. *Sulphate of Lead*, Vitriol of Lead; found in

in octohedrons, and their modifications; of a yellowish brown colour; matrix upon a ferruginous quartz or schistus. Found in Anglesea, Wales; and near Strontian, Scotland.

n. *Phosphorated Lead Ore*; lead combined with phosphoric acid; shining, semi-transparent; found near Frieberg, Zschoapau, Johanngeorgenstadt, Zellerfeld, and in the lead hills, Scotland.

o. *Yellow Lead Ore*; or lead combined with acid of Tungsten; yellow or orange coloured lamellated, semi transparent; is the molybdate of lead of Klaproth, who found it combined with the molybdenic acid. It is found at Bleiberg in Carinthia, near Villach, near Zellerfeld, and in the lead hills in Scotland.

F. COPPER.

COPPER, when pure, is near nine times heavier than water; the most sonorous of all metals, dissolves in all acids and alkaline solutions, oils and water. The least quantity of this metal, in solution, turns blue by the addition of volatile alkali: united with calamine, it forms brass; with tin, bell-metal.

a. *Native Copper*; malleable, fibrous, fusible, generally found adhering to other fossil substances; crystallized in cubes, octohedrons, compact, laminar, granular.

b. *Native*

b. *Native Oxyde of Copper;* found in compact lumps, ſprinkled in the ſtate of ſmall particles, of a hyacinth colour. Sometimes reſembling brown pitch, or of a ſteel grey, and browniſh red. Cryſtallized in cubes and octohedrons.

c. *Pitch Copper Ore*, hyacinth red; little ſhining, ſoft, compact, pulverulent; containing oxyde of copper with oxyde of iron *Born.*

d. *Carbonate of Copper*; Malachite, ſky blue, moſtly pulverulent, cryſtallized in rectangular octohedrons and various; ſtalactitical. Found in Tyrol. *Anal.* Copper 73. Carbonic acid 26. *Fontana.*

e. *Arseniate of Copper*; of a deep olive, or emerald colour, ſemi-tranſparent, cryſtallized in various modifications. *Anal.* Copper 73. Carbonic acid 26. *Fontana.*

	Synonymous Names.
f. *Sulphate of Copper*,	Vitriol of Cyprus. Blue Vitriol. Vitriol of copper, or of Venus. Blue copperas.

cryſtallized in tetrahedral priſms, and their modifications, ſhining, colour braſs yellow: contains copper, iron, and ſulphur.

g. *Muriate** *of Copper*; amorphous, in a ſandy

* Salt formed by the union of the muriatic acid with different baſes.

form.

form. *Anal.* Copper 52. Acid 10. Oxygen 11. Water 12. Sand 11. *Rochefoucauld.*

h. *Sulphuret of Copper,* Vitreous Copper Ore, or Pyrites of Copper; of a lead grey colour, cryſtallized in hexahedral truncated priſms; contains copper with ſulphur: ſometimes a little ſilver or arſenic. Found at Freiberg.

i. *Variegated Copper Ore;* exhibiting various colours; intermixed with copper pyrites and vitreous copper. Found at Freiberg.

k. *Yellow Copper Ore;* cryſtallized in equilateral tetrahedrons, and various; compoſed of copper, iron and ſulphur.

l. *Grey Copper Ore;* cryſtallized in tetrahedrons and their modifications, and in ſix-ſided priſms. *Anal.* Copper 16. Lead 34. Antimony 16. Iron 13. Sulphur 10. Silver 2. Silex 2. *Klaproth.*

m. *White Copper Ore;* ſilver white, brittle; on rubbing emits an arſenical ſmell; compounded of copper, iron, and arſenic. Found near Freiberg.

G. IRON.

IRON is about eight times heavier than water, is attracted by one of it's ores, called the loadſtone; ſoluble in all acids, alkaline ſolutions, water, and air; it's ſolution is turned of a black or dark purple colour, by galls, and other vegetable aſtringents.

a. *Native Iron*: various, irregular, malleable; of an iron grey colour; found at Kamſdorf, in Siberia; and South America.

b. *Grey Iron Ore*, Magnetic Iron Stone, Emery; ſhining, copper red colour, a little attracted by the magnet; cryſtallized in octohedrons, and in cubes; lamellated, foliated, granular, and various; contains iron united to a ſmall proportion of oxygen.

c. *Hematite*; of a blood red, or ſteel grey colour; found ſolid, reniform, fibrous, ſcaly and various; found at Schneeberg, &c.; contains iron with carbonic acid and argill. *Kirwan.*

d. *Argillaceous Iron Ore*, Baſaltic Iron Ore; reniform, globular, lenticular, granulated, various; found at Bohnerz in Germany; contains iron united to oxygen, carbonic acid and argill, and often phoſphate of iron.

e. *Spathose Iron Ore*; of a greeniſh grey cream colour, or chocolate brown; cryſtallized in rhombs, and in double, three, and ſix-ſided pyramids, foliated; found near Freiberg, Kamſdorf, &c. in Germany. *Anal.* Iron 38. Lime 38. Carbonic acid and manganeſe 24. *Bergman.*

	Synonymous Names.
f. *Sulphate of Iron.*	Martial vitriol. Green vitriol. Vitriol of iron. Green copperas.

a combination of the ſulphuric acid with iron.

g. *Sulphuret of Iron,* Martial Pyrites; cryſtallized in tetrahedrons, and their modifications, in cubes, octohedrons, ſolid, interſperſed, capillary, ſtriated; conſiſts of iron and ſulphur.

Mr. David Muſchet, of the Clyde iron works, divides iron-ſtones into,

1. Argillaceous Iron-ſtone, having clay for it's chief component earth, and this clay comparatively pure and free from ſand.

2. Calcareous Iron-ſtone, poſſeſſing lime for it's chief mixture, and this lime alſo comparatively deſtitute of ſand.

3. Siliceous Iron-ſtone, uniting clay and lime, containing large proportions of ſilex.

4. Iron-ſtone containing nearly the ſame proportions of clay, lime and ſilex.

Beſides theſe are deſcribed, " Primary ores of iron," ſo named in contradiſtinction to ores which appear, like iron-ſtones, to have been formed by a ſecondary agency.

H. TIN.

H. TIN.

TIN is a ſilver-coloured gliſtering metal, ſeven times heavier than water; does not vitrify like lead; is malleable and unſonorous; ſoluble in aqua regia, vitriolic, muriatic, nitric and acetous acids; yields a white oxyde that impairs the tranſparency of glaſs, and converts it into enamel. Found in Cornwall and Bohemia.

a. *Native Tin:* pure native tin is ſo very rare, that it's real exiſtence has been doubted.

(a.) 2. *Native Oxyde*, Spathoſe tin ore; tin combined with oxygen and iron; is found in various modifications.

b. *Brown Tin-stone and Spar;* conſiſts of calx of tin, calx of iron, and acid of tungſten.

c. *Wood Tin*, Stream Tin, or Corniſh Tin Ore; found only in Cornwall, in ſmall globular or reniform pieces, ſometimes of a fibrous or radiated texture; containing tin, with oxygen and iron.

d *Tin Pyrites*, Sulphuriſed Tin, or Sulphuret of Tin; contains tin, ſulphur, copper and iron, beſides it's matrix; it is diſtinguiſhed by it's ſulphureous ſmell when heated.

II. Fra-

II. Fragile.

A. BISMUTH.

BISMUTH is of a white yellowish colour, and laminated texture; above nine times heavier than water, very fusible, soluble in the sulphuric muriatic and nitric acids, and particularly in the last, and when dissolved in it, is precipitable by a mere dilution with pure water; the precipitate is white. Found plentifully in Saxony, Bohemia, Sweden.

a. *Native Bismuth*, of a silver white colour, inclining to reddish. It's matrix is red jasper, petrosilex, quartz, heavy spar, and cobalt ores. Found crystallized in equilateral triangular laminæ; indeterminate; in small particles, foliated. Separated from it's matrix by simple fusion; being itself easily fusible.

b. *Native Calx, or Oxyde of Bismuth;* without lustre; friable, loose, or compact; of an earthy appearance; emitting no sulphureous smell when ignited; soluble in nitric acid; of a yellowish grey, greenish, and straw yellow grey.

c. *Sulphurised Bismuth*, or Sulphuret of Bismuth; of a lead grey and bluish grey colour; sometimes variegated; amorphous, striated, foliated, loose and dispersed. *Anal.* Bismuth 60. Sulphur 40. *Sage.*

d. *Martial*

d. *Martial sulphurated Bismuth.* Of a radiated texture; yellowiſh grey colour. Found near Gillabet in Norway.

e. *Arsenicated Bismuthic Ore;* of a yellowiſh white colour; brilliant luſtre; emits a garlick ſmell when ignited; conſiſts of biſmuth, arſenic, and ſulphur. Found at Schneeberg in ferruginous jaſper, accompanied with cobalt ore.

B. NICKEL.

Found generally in a metallic ſtate, more rarely in the ſtate of calx. When free from heterogeneous ſubſtances, is of a grey reddiſh white, or fleſh colour; when broken, has a ſtrong luſtre; of a fine-grained, compact texture; has a little ductility; is ſoluble in ſulphuric, muriatic, and particularly nitric acid.

a. *Native Nickel;* copper-red colour; exhibits a conchoidal ſurface when broken; cryſtallized in rhomboidal tables; amorphous, foliated. It's matrix is calcareous ſpar, and heavy ſpar. *Anal.* Nickel alloyed by iron. *Born.*

b. *Native Calx or Oxyde of Nickel;* colour of a pale green, or bluiſh green; has an earthy appearance; is friable; ſometimes ſhining. *Klaproth* found it mixed with ſiliceous, argillaceous, magneſian, and calcareous earth.

c. *Kupfer Nickel,* copper-red colour; deeper than

than the purified metal; of particular ſhapes; amorphous, granular, compact. *Anal.* Nickel united to iron, arſenic, cobalt, and ſulphur. *Bergman.*

C. ARSENIC.

ARSENIC is about eight times heavier than water, appears in plates of a bluiſh grey colour, may be diſcovered by evaporating it upon red-hot coal or iron, by means of which a garlic ſmell is emitted. The fumes depoſit a white coating on a plate of copper. It is not acted upon by water, but readily by nitric acid. Phlogiſticated alkali dropped into it's ſolution, produces a green precipitate.

a. *Native Arsenic*, Teſtaceous Arſenic; arſenic alloyed by iron. It is generally found in cobalt mines, in Saxony, Bohemia, Norway.

b. *Native Oxyde of Arsenic*, White Oxyde of Arſenic; this is ſcarce, of a whitiſh colour, but expoſed to heat becomes blackiſh. It is ſoluble in water aſſiſted by heat; alſo in nitric acid, leſs ſo in muriatic and vitriolic acid. Found in Hungary, on the Hartz, in Bohemia, and Tranſylvania.

c. *Sulphuret of Arsenic*, Sulphuriſed Arſenic, Realgar, yellow Orpiment; of a yellow brimſtone colour, inclining to orange; it's texture uſually lamellar, and ſo ſoft as to be cut with a knife, and a little a little flexible; often of a beautiful brilliant lustre, which is improved by the ſcarlet red ſhades. When reduced to powder, it is uſed as a pigment. Found in Hungary, and other places. *Anal.* Arſenic 84,90. Sulphur 16,10. *Kirwan.*

d. *Red Sulphurized Arsenic*, or Ruby Arſenic; of an aurora colour; contains a larger proportion of ſulphur than the preceding; common in China, where it is made into vaſes, pagodas and other ornamental works. Red arſenic is commonly found near volcanos, as at the Solfatara near Naples, &c.; it's matrix is quartz, heavy ſpar and ferruginous clay.

e. *Mispickel*, Pyritical Arſenical Ore, Arſenical Mundick; arſenic and iron mineraliſed by ſulphur; ſometimes combined with ſilver; it is found in cubes, rhomboidal, four-ſided, truncated priſms, ſometimes terminating in dihedral ſummits, with triangular plans; octohedral; the ſurface is generally ſtriated. When encloſed in a cloſe veſſel, it ſublimes and forms orpiment, leaving the iron behind. It's matrix is ſpathoſe iron ore, fluor, quartz, blende. Found in Saxony, Bohemia, Tuſcany, and on the Hartz.

D. COBALT.

COBALT is ſeven times heavier than water, of a roſe-white colour, brittle and eaſily reducible to powder; is a little moveable by the magnet, probably from ſome iron it contains; difficult of fuſion; when fuſed with ſand and potaſh, it forms ſmalts,

ſmalts, uſed for blueing clothes. Is ſoluble in nitric acid without heat, and in vitriolic acid aſſiſted by heat; likewiſe in the marine and nitro-muriatic acid, which diluted with water, is uſed for ſympathetic ink.

a. *Grey Cobalt Ore,* Arſenicated Cobalt; chiefly compoſed of cobalt and arſenic; of a fine-grained, compact texture; of a ſteel-grey colour. Found in Saxony, Bohemia, Norway, in a matrix of red heavy ſpar; calcareous quartz.

b. *White Cobalt Ore,* conſiſts of cobalt and iron, mineraliſed by ſulphur and arſenic, of a tin-white colour, of a granular and lamellated texture. Found in Norway and Sweden.

c. *Sulphuret of Cobalt*; contains no arſenic or iron; when heated it emits a ſulphureous ſmell; found in cubical cryſtals in Upper Hungary, generally upon quartz.

d. *Native Oxyde of Cobalt*; Black Oxyde or Calx of Cobalt; this is the pureſt kind of the calciform cobalt ores; when heated, it emits no ſulphureous or arſenical vapours. Found in Thuringia, Saxony, Tyrol. Of the oxydes of cobalt there are various ſpecies, as *Brown earthy Oxyde, Yellow Oxyde, Green Oxyde, Red Oxyde, of Cobalt.*

N. ZINC.

ZINC is about 7 times heavier than water; it is ſoluble in the nitric, ſulphuric, and muriatic acids, and produces by the two laſt hydrogen gas or inflammable air; it burns with a bluiſh green flame when expoſed to red heat, and ſublimes in the ſtate of a white light ſubſtance; it precipitates lead and other metals, in a metallic ſtate, from their ſolution. When in an oxyde ſtate, is recovered by charcoal in a cloſe veſſel. It is found in various parts of Europe, never in a perfect metallic ſtate, but generally in the ſtate of calx or oxydated; frequently alſo combined with iron and ſulphur. It is of a fibrous texture, and of a bluiſh grey colour.

a. *Native Oxyde of Zinc*, calamine, calx of zinc. *Anal.* Oxyde of zinc, 84. Oxyde of iron 3. Silex 12. Argill 1. *Bergman.*

	Synonymous Names.
b. *Carbonate of Zinc*,	Chalk of Zinc.
	Aërated Zinc.
	Mephite of Zinc.

Oxyde or Calx of Zinc, combined with, or mineraliſed by fixed air or carbonic acid. Is perfectly ſoluble in ſulphuric acid, without emitting heat; the fixed air is diſcovered by the efferveſcence which takes place when diſſolved in the ſulphuric acid, and the zinc combining with the acid, forms

c. *Sulphate*

Synonymous Names.

c. *Sulphate of Zinc,* { Vitriol of Zinc. White Vitriol. Vitriol of Goslar. White Copperas.

soluble in water; formed from blende, acidified by the absorption of oxygen. and thus producing the sulphuric acid, which afterwards combining with the calx of zinc, composes the Sulphate or Vitriol of Zinc. *Anal.* Zinc 20. Sulphuric acid 40. Water 40. *Bergman.*

d. *Blende,* or Sulphurized Zinc; combined with sulphur, iron, copper, silex. It is of various colours, sometimes resembling galena, and hence called pseudo-galena; when of a blackish colour, it is called *blackjack. Blende,* in the German, signifies blinding or deceitful. *Anal.* Zinc 52. Sulphur 26. Iron 8. Copper 4. Silex 6. Water 4. *Bergman.*

O. ANTIMONY.

ANTIMONY is about 7 times heavier than water; soluble in aqua regia, and precipitated by pure water; scarcely soluble in sulphuric or nitric acid; with muriatic acid it forms butter of antimony.

It is found in nature in the metallic state; in the state of calx, mineralised by arsenic, but most generally with sulphur, called crude antimony. It is of a bluish colour, of a lamellated and radiated texture.

a. *Native Antimony*, usually mixed with iron and arsenic. It contains so large a proportion of the latter, that by fusion with sulphur, the product resembles realgar, or red arsenic; found in Dauphiny, Sweden, Saxony.

b. *Arsenicated Antimony*, Native arsenical Antimony; of a white brilliant lustre, and scaly texture; when heated, it emits only arsenical vapours. *Anal.* Antimony alloyed by arsenic.

c. *Muriate of Antimony*, White antimonial Ore; combined with muriatic acid, of a greyish white colour, found in oblong, rectangular, four-sided laminæ. It is scarce; found near Braunsdorf, Freiberg, Bohemia; generally accompanied by red blende, galena, grey antimony.

d. *Red Antimonial Ore*; consisting of antimony, sulphur, and arsenic; of a light crimson red colour, in hexahedral, prismatic, capillary crystals; found in the antimonial mines of Bohemia, Hungary, and Transylvania.

e. *Sulphuret of Antimony*; Sulphurised Antimony; of a beautiful appearance of the rainbow colours; filaments long, in hexahedral prisms, with obtuse tetrahedral prisms; it is extremely scarce; it was found in a certain antimonial mine of Hungary, in quartz and heavy spar. *Anal.* Antimony 74. Sulphur 26. *Bergman.*

f. *Plumose Antimonial Ore*, or Argentiferous antimonial

antimonial Ore; antimony combined with iron, arsenic, sulphur, and sometimes silver; of a lead-grey colour; crystallized in slender prismatic needles. It is found near Freiberg, in Hungary and Tuscany.

P. MANGANESE.

MANGANESE is about 7 times heavier than water; soluble in marine acid; the solution appears brown; by addition of marine acid to manganese, is produced an elastic fluid or gas, called suroxygenated muriatic gas, or dephlogisticated muriatic acid air. When exposed to red heat in a close vessel, pure air is produced. It is usually found in the state of calx, or combined with oxygen. Next to iron and gold, it is the most frequently diffused metallic substance through the earth, and even in vegetables. Soluble in acids, and precipitated in a white powder by alkali.

a. *Native Manganese*; found in the state of greyish-white small globules, which, on exposure to the atmosphere, fall to a black powder.

b. *Native Oxyde of Manganese*, White oxydated Manganese; crystallized in tetrahedral rhomboidal prisms; found in Hungary, Transylvania, mixed with silex, and constitutes the matrix of the auriferous ores. *Anal.* Oxyde of Manganese 43. Oxyde of Iron 43. Lead 4½. Mica 5. *Wedgewood.*

c. *Black*

c. *Black calciform Manganese*, Black-Wad; of a greyiſh, or iron-black colour; found in Könitz, Piedmont, Carinthia, Thuringia, England, and Ireland.

d. *Siliceous Ore of Manganese*; of a ſteel-grey colour, and lamellated texture; combined with oxyde of mangaueſe 35. Silex 55. Iron 5; and argill 5.

Q. SCHEELE.

SCHEELE, *Wolfram*, or *Tungsten*; of a greyiſh white, or yellowiſh colour; granulated, friable; is never found in the metallic ſtate, but alone in the ſtate of calx, combined chiefly with calcareous earth, manganeſe and iron; it's oxyde is of an acid nature, of a yellow colour, and enters into combination with all metals; it is ſoluble in the ſulphuric, nitric, muriatic, and nitro-muriatic acids, and is converted by them into an oxyde, which combines alſo with alkalis, but is precipitated again from them by nitric acid. With the muriatic acid, the oxyde aſſumes a blue colour.

a. *Tunstate of Lime**, Sparry Tunſtate of Lime, Tungſten; combined with Tungſtenic acid 44; and lime 56; of a lamellar texture; greyiſh white colour; cryſtallized in angular octohedrons, or double tetrahedral pyramids; is 6 times hea-

* Salt formed by the combination of the Tunſtic acid with different baſes.

vier

vier than water; digested with muriatic acid, it turns yellow; found in Bohemia and the tin mines in Cornwall,

b. *Wolfram*, Manganeseous Wolfram; combined with tungstenic acid 64; oxyde of manganese 22; oxyde of iron 13; silex and tin 2; of a dark black colour; lamellated, brittle, crystallized in compressed hexahedral prisms, terminated with tetrahedral pyramids; found in England, Siberia, Bohemia, Saxony.

R. URANITE.

URANITE, or Uranium; is of a deep-grey colour, with a slight lustre; above 6 times heavier than water; is soft enough to be cut with a knife; soluble in nitric, sulphuric, and marine acids; is not precipitated by zinc; phlogisticated alkali added to the solution, produces a deep red precipitate. First separated in 1790, by Klaproth, from the mineral called Pechblende. Found in a state of calx or oxyde, and mineralised by sulphur.

a. *Carbonate of Uranite*; Calciform or Oxyde of Uranite, or Calcolite. The earthy oxyde of uranite is a variety of an earthy texture; of a brimstone yellow; it is found solid, dispersed through, and deposited upon other ores. The variety called spathose uranite, is of a grass-green colour, and generally found crystallized in cubes, and in hexahedral prisms; consists of uranium, carbonic

carbonic acid, and a little copper; found in Bohemia.

b. *Sulphuret of Uranite*; Sulphurated Uranite, Pechblende; of an iron colour, brittle texture; emits a ſulphureous ſmell expoſed to heat.

S. MOLYBDENA.

MOLYBDENA, is of a ſteel-grey colour, very brittle; 6 times heavier than water; is only found in nature combined with ſulphur. It melts eaſily; is volatile by moderate heat, efferveſces with alkali; emits a ſulphureous ſmell when treated with the blowpipe.

a. *Sulphurised Molybdena*, or Molybdena mineraliſed by ſulphur; conſiſting of the molybdic acid and ſulphur; it's matrix is feltſpar, lithomarge, and quartz; found in Iceland, Sweden, Spain, Saxony, France, Siberia; frequently in rocks containing wolfram and tin ores.

T. MENACHANITE.

MENACHANITE, diſcovered in the valley of Menachan, in Cornwall, by Mr. Gregor; and ſuppoſed to conſtitute a new metallic ſubſtance.

a. *Native Menachanite*; Menachanite alloyed by iron; found amorphous, or in grains, and in irregular forms; of a grey, or dark colour; ponderous, and attractable by the magnet.

U. SYL-

U. SYLVANITE.

This ſemi-metal, diſcovered by *Facebay*, is different from all other known metallic ſubſtances; and is called Sylvanite from it's being found in Tranſylvania; of a metallic luſtre; fracture broad, or granularly imbedded; of a dark-grey or white colour; reſembles regulus of antimony, or ſulphurated biſmuth; evaporates by continued heat; amalgamates with mercury by ſimple trituration; and detonates with nitre; ſoluble in aqua regia, and in concentrated vitriolic acid in cold, or a low digeſting heat, and the ſolution is crimſon red; but by the affuſion of water, or by a ſtrong heat it is precipitated. Muller found it to contain a ſmall proportion of arſenic and of nickel, and alſo of gold. Bergman found it to contain a little zinc; but theſe mixtures ſeem to be merely caſual. *Kirwan.*

V. TITANITE.

A ſemi-metal, diſcovered by Klaproth, found at Rhonitz, and at Bainick in Hungary; of a browniſh-red colour; fracture foliated, unequal and conchoidal; cryſtallized in right-angled quadrangular priſms, longitudinally ſtreaked and furrowed, often acicular, and ſeated on ſchiſtoſe mica alternating with quartz. Expoſed to a moderate heat, it remains unaltered, but in a cool crucible it burſts into angular fragments, loſes it's luſtre and colour, and becomes pale brown; inſoluble in aqua regia, vitriolic, nitrous, or marine acid.

a. *Calcareo Siliceous Ore*; found near Paſſau; diſcovered and deſcribed by Profeſſor Hunger; of a reddiſh, yellowiſh, or blackiſh-brown colour; found maſſive or diſſeminated; but more frequently cryſtallized in obtuſe-angled tetrahedral cryſtals; found ſeated on, or inhering in gneiſs or granite; foliated or ſtriated; cuts glaſs. *Kirwan. Anal.* Silex 35. Calcareous earth 33. Titanitic calx 33. *Klaproth.*

W. CHROME.

CHROME, ſo called from it's property of colouring the combinations into which it enters; diſcovered by Klaproth in the red lead of Siberia. Vauquelin has made many experiments upon this ſubſtance, but it's place in mineralogy is not yet fully eſtabliſhed.

X. TELLURIUM.

Klaproth has chemically analyzed the auriferous mine, known as the mine of white gold (Weiſe goldertz) aurum paradoxum, metallum vel aurum problematicum; found in the mine called Maria-lulf, in the mountain of Fatzbay near Zalethna, in Tranſylvania; and in this mineral has found a metal different from every other yet known.

It is of a whitiſh greyiſh colour; friable and lamellar; in cooling after being heated, it takes a cryſtallized ſurface; very hard of fuſion. Placed on a heated body, it burns with a bright flame, of a blue colour; it amalgamates with mercury.

It's

It's ſolution in the nitrous acid is clear and colourleſs; when it is concentrated, it exhibits pointed needle-like cryſtals.

The pure alkali precipitates from the acid ſolutions of it, an oxyde of a white colour, ſoluble in all the acids.

The ore of white gold of Fatzbay, aurum vel metallum problematicum, contains: *Anal.* Metal of Tellurium 925,5. Iron 72,0. Gold 2,5. Total 1000.

The graphic gold of Offenbanya, contains Tellurium 60. Gold 30. Silver 10. Total 100.

The mineral known under the name of the *Yellow Ore of Nagyag*, contains Tellurium 45. Gold 27. Lead 19,5. Silver 8,5. Sulphur ſcarcely an atom. Total 100,0.

The *Foliated grey Gold of Nagyag*, contains Lead 50. Tellurium 33. Silver 8,5. Sulphur 7,5. Silver and Copper 1. Total 100,0.

CLASS IV.

INFLAMMABLES.

INFLAMMATION, or combuſtion, conſiſts in the fixation and abſorption of vital air by combuſtible bodies, and in the decompoſition of atmoſpheric air by theſe bodies. As vital air is a gas, and many combuſtible bodies, by abſorbing, fix it, and make it take a ſolid form; hence vital air, when thus precipitated, will loſe the caloric to which it owed it's elaſtic fluidity; and hence comes that free caloric or heat, which is evolved during combuſtion.

I. Aeriform.

A. *Hydrogen*,

Is one of the component principles of water; and in combination with caloric and light, forms hydrogen gas.

a. *Pure Hydrogen*, exiſts always in a ſtate of gas. It is a conſtituent of water; where it forms one ſixth. It is diſtinguiſhed from all other gaſes, on on account of it's being inflammable; hence it is called inflammable air. See §. V. page

b. *Sulphurated Hydrogen*, *Hepatic Gas*, is obtained from livers of ſulphur, or ſulphures, by decompoſing them with acids. Sulphur combines with hydrogen gas, called ſulphuret, or hydrogen gas, or ſulphurated hydrogen gas, or hepatic air.

II. Simple bituminous Substances.

Bituminous ſubſtances are not of mineral origin, but have been formed from certain principles of ſubſtances belonging to the organiſed kingdoms of nature; which, after the loſs of animal and vegetable life, have ſuffered conſiderable changes, by long contact and union with mineral bodies; and may hence be regarded as forming a part of the mineral ſyſtem.

a. *Naptha*, is a light, thin, often colourleſs oil, highly odoriferous and inflammable, which in ſome parts of Perſia and Italy is found upon the ſurface of ſprings and lakes, and iſſuing from argillaceous ſtones; does not combine with water or ſpirit of wine; when burned, leaves a black ſoot; acids condenſe it and render it reſinous: when firſt expoſed to the air, it becomes yellow; afterwards, and in like proportion, it thickens, and paſſes into

b. *Petroleum, or Rock Oil;* and ſeems chiefly to differ from Naptha in being mixed with heterogeneous

geneous matter; it is found in various parts of Europe trickling from the fissures of rocks. It has a greasy feel, is transparent, or semi-transparent, of a reddish or blackish-brown colour; by air it becomes like tar, and then is called

c. *Mountain or Mineral Tar, Barbadoes Tar,* Bitumen Petroleum tardefluens. It is viscid; of a reddish or blackish-brown or black; when burned it emits a disagreeable bituminous smell, and by exposure to the air, it passes into

d. *Mountain or Mineral Pitch, Bitumen Maltha;* which is more impure than petroleum; it has the consistence of honey; is of a brownish-black colour; frequently mixed with much earthy matter, resembles common pitch; when heated emits a strong unpleasant odour, like the former substance; in cold weather it may be broken, and exhibits internally a glossy lustre; when warmed it softens and possesses some tenacity. It is however susceptible of a superior degree of induration, and then becomes *asphaltum.*

e. *Asphaltum,* or *Bitumen of Judea,* has derived it's name from the Dead Sea of Judea, upon the surface of which it is found, and upon the shore. It is met with also in other countries. It is smooth, of a black colour, and shining conchoidal fracture.

III. COM-

III. Compound bituminous Substances.

Mr. Hatchett, whofe divifion of bituminous fubftances I have chiefly followed, is of opinion that the progreffive changes of naptha into petroleum, mineral tar, mineral pitch, and afphaltum, is caufed by the gradual diffipation of part of the hydrogen of the bitumen, and the confequent difengagement of carbon; and although the divifion of the fimple bituminous fubftances terminates in afphaltum, nature appears to have glided on by an uninterrupted chain, which connects the fimple bitumens with the compounds.

a. *Jet*, is a compact black body; harder and lefs brittle than afphaltum; may be formed into trinkets; it breaks with a conchoidal fracture, and the internal luftre is gloffy. It has no odour, unlefs when heated, and it then refembles afphaltum.

b. *Coal*, of which carbon is a conftituent principle, is found in various parts of Europe in ftrata beneath the furface of the earth; it is foliated, brittle, black, and glittering. Befides carbon, it's conftituent parts are, petroleum, maltha, and afphaltum, combined more or lefs with other earthy particles, and, frequently with fulphur, and the remains of vegetable matter.

c. *Mineral Elastic Gum;* difcovered in 1786, near

near Caſtleton, in Derbyſhire; reſembles, in colour and elaſticity, cahout-chou, or Indian rubber; formed from naptha or petroleum; there are many varieties; ſome in the ſtate of aſphaltum, by melting loſe their elaſtic property, and a quantity of gas is diſengaged; and the ſubſtance remaining reſembles mineral pitch, or mineral tar.

B. AMBER is one of the pureſt bitumens, conſiſting almoſt entirely of a volatile oil, which receives it's conſiſtence from a peculiar acid (ſuccinic) and of a few earthy and coaly particles.

Amber has been chiefly procured on the ſhores of the Baltic. The yellow amber of Kœnigſberg has been lately diſcovered by digging. It is raiſed 200 feet from the ſea, where many ſhafts are ſunk, one of which is near 100 feet in depth. The amber does not run in veins, but is found in nodules, in a matrix of charcoal, below which there are ſtrata of ſand. In the charcoal there are ſometimes little threads of amber. This ſeems to ſhew that the origin was volcanic, and that the amber was accumulated by ſublimation.

a. *Honeystone;* cryſtallized in aluminiform octohedrals; found in Thuringia, between beds of bituminous wood, and is ſometimes near an inch in diameter; it is cryſtallized in great maſſes, in a double pyramidal form, with four ſides; it bears a great reſemblance to amber, but does not, like that ſubſtance, become electric by rubbing; by Mr. Abich's analyſis it appears, that 50 parts of this foſſil are compoſed of 8 parts of the carbonate

of

of alum, 2 of carbon, 1½ of oxyde of iron, 20 of carbonic acid, 14 of the water of cryſtallization, emitting the ſmell of bitter almonds, and 2¾ of ether; with a ſmall portion of the benzoic acid. In conſequence of the incombuſtibility of the honey-ſtone, he propoſes to exclude it from the claſs of combuſtible bodies, and to place it among that of aluminous earths.

b. *Common Amber* is found in ſeveral countries of Europe beneath the ſurface of the earth, among clay, ſand, and the iron bog-ore, but moſt abundantly in the ſea, and particularly on the ſhore of the Baltic; found in maſſes, irregular, tranſparent or opaque, of a white, yellow, or brown colour; red and green amber being ſcarce. Taſteleſs; of a faint odour when rubbed, and becomes electric; inſoluble in water.

C. MINERAL TALLOW, or MUMIA; found in 1736, in the ſea, on the coaſt of Finland; alſo in ſome rocky parts of Perſia, mixed with petroleum. It is perfectly white, of the conſiſtence of tallow, but more brittle though as greaſy; it burns with a blue flame, and a ſmell of greaſe, leaving a black viſcid matter, which is more difficultly conſumed.

D. SULPHUR is a ſimple body, which, during combuſtion, or, in other words, during it's combination with oxygen, produces ſulphuric acid. Sulphur is found near the craters of volcanos, and the ſources of ſome mineral waters.

a. *Common, or Native Sulphur*, is a yellowish substance, odorate, electric, transparent, and octohedral; opaque and prismatic; fusible, and subject to different combustions; one is slow, with a bluish flame, and yields sulphureous acid; the other rapid, with a white flame, and yields sulphuric acid.

b. *Volcanic Sulphur;* amorphous, granular, compact.

E. PLUMBAGO, or *Carbure of Iron*; or iron combined with a large portion of carbon.

a. *Common Plumbago; Graphite;* this substance has not yet been produced by art, but nature contains it in considerable abundance; suffers no change by water or atmospheric air, nor exposed to fire in close vessels; but burns slowly, if heated, in contact with air; amorphous, fibrous and slaty; of a deep iron black colour; little shining, metallic lustre, stains the fingers, soapy to the touch.

b. *Anthracolith;* consisting of Carbon 90. Argill 5: Iron 3. *Born.*

VOLCANIC PRODUCTIONS.

I. CINDERS.

A. Loose.

a. *Ashes*, of a browniſh or reddiſh grey colour; conſiſtence looſe and duſty; very light and ſubtle, and ſmooth to the touch; ſlowly diffuſible in water, and when wet, ſomewhat ductile; effervesce ſlightly with acids; and contain about half their weight of argill, a ſmall proportion of earth, calx, magneſia, and iron, and the remainder is ſiliceous.

b. *Sand;* conſiſting of minute hard grains, that readily ſink in water; generally fragments of lava and ſcoriæ, together with olivia, garnets, feltſpar, and ſchorl.

B. Coherent.

a. *Puzzolana;* colour reddiſh brown, grey, or greyiſh black; that of Naples grey; that of Civita Vecchia, reddiſh brown; ſurface rough, uneven, and of a baked appearance; pieces of various ſizes, from the ſize of a nut to that of an egg; fracture uneven, or earthy, and porous; commonly filled with particles of pumice, quartz, scoriæ, &c.; very brittle; does not effervesce witl acids; magnetic before it is heated, but not after; contains silex, argill, and iron.

b. *Trass*, or *Terras*; resembles puzzolana; though often the product of pseudo volcanos, or external fires; colour greyish brown, or yellowish: surface rough and porous; fracture earthy, rarely lamellar; found principally near Andernach, in the vicinity of the Rhine; also near Frankfort, Cologne, Pleith, &c.; and there called Duffstein; containing fragments of argillite and basaltine, often branches of trees half cleared, and impressions of leaves, with mica, iron ore, &c.

c. *Tasa*; a kind of puzzolana formed by nature into a mortar; colour brown, or reddish brown, brick red, or speckled with various colours; fracture earthy; contains sand, scoriæ, fragments of lava, limestone and pumice; often basaltines and Vesuvian; commonly magnetic, and not easily decomposed by the action of the ore.

d. *Pumice;* colour grey, rarely brown, or blackened by fuliginous fumes; surface rough and fibrous, with elongated pores; sometimes fibres not discernible; fracture striated and open, uneven and splintery; fragments oblong, obtuse, and irregular; brittle in a high degree; fusible at 130. o. into a grey flagg.

e. *Piperino;*

e. *Piperino*; a concretion of volcanic ashes, and is said to be the substance that covers Pompeia; colour grey or reddish brown; fracture earthy; contains fragments of white marble, feltspar, mica, garnets, scoriæ, gypsum, schorl, granite, &c. sometimes magnetic; if preserved from moisture, it hardens in the air.

II. LAVA.

Any matter that has issued out of a volcano in a liquified state, or that has accompanied or been enveloped in such liquified matter. *Kirwan.*

a. *Cellular;* colour brown, or greyish black; surface unequal, rough, and full of cavities; affects the magnetic needle.

b. *Compact;* earthy substance, or matter, which, after having been fused, but not vitrified, becomes, on cooling, compact, close, and solid.

III. VITREOUS LAVAS.

a. *Glass*; such matter as has been exposed to the greatest heat, and from the composition, most susceptible of vitrification in a moderate heat; colour black, green, yellow, or greyish white; found in detached masses; either spongy or compact fracture.

b. *Enamel*; resembling vitreous lavas, is the imperfect vitrification called enamel; colour white,

white, grey, or brown, often with round ſpots, of a different colour from the ground; ſometimes contains cryſtals of ſchorl, or feltſpar, not completely melted; fracture rather more granular than that of glaſs.

c. *Scoriæ*; compounded of iron and ſtony matter, and owe their fuſion to ſulphurated iron; they are tumefied and expanded by the ſulphureous vapours; colour generally black; ſometimes reddiſh from the calcination of the iron; ſurface rough, and uneven; texture cavernous; the cavities larger and more irregular than thoſe of cellular lava, but never fibrous.

d. *Slaggs*, reſemble the droſs of forges, generally red and heavy.

THE

NATURALIST's and TRAVELLER's COMPANION, &c.

PART the SECOND.

ALTHOUGH it may he admitted with peculiar honour to the present age, that the knowledge of natural history and of science in general has been of late considerably enlarged; yet as the objects of human inquiry are numberless, and frequently dispersed in distant parts of the globe, as well as complicated in their history, the sentiments of an ancient philosopher may be adopted even at this day with propriety: "Multa etenim "sunt quæ esse audivimus, qualia autem sint ig-"noremus! Quamque multa venientis ævi popu-"lus, ignota nobis, sciet*!"

At the same time if we reflect upon the foregoing suggestion, respecting the amazing progress made in natural history within the space of a few years, we may find sufficient inducement to persevere in pursuits so worthy of a rational mind.

* Seneca.

It

It would render natural hiſtory much more pleaſing, as well as greatly tend to it's progreſs, were the limits of our knowledge therein preciſely aſcertained, that travellers and curious perſons, who have little leiſure for reading, might not only be informed of what is already diſcovered, but alſo of what is ſtill doubted, or unknown; by which means their inquiries would be better directed, and more conducive to real information and uſeful diſcovery.

SECT.

SECT I.

Observations and Queries respecting Learning, Antiquities, religious Rites, polite Arts, &c.

Ingenuous arts, where they an entrance find,
Soften the manners, and ſubdue the mind*.

1. THE alphabets of the various nations, their pronunciation and numeric value, with their numeric figures, if different from the letters of the alphabet, and books written in each language, eſpecially grammars, dictionaries, &c. with the dates of each when written, merit the inveſtigation of the curious; likewiſe the materials uſed for writing, and their preparation, as the methods of making ink, paper, and pens, and of ſizing and gluing the paper; the art of printing, and the contrivances for doing it.

2. Manuſcripts, in good preſervation, of the Hebrew Bible, or parts thereof, particularly if upwards of 300 years old.

3. Books containing the religious principles of any nation or people, and which uſually are written in a dialect different from that which ſuch people now ſpeak, or in a poetical, high metaphorical ſtyle, and therefore underſtood by few only, and for the moſt part kept very ſecret;

* ——— Ingenuas didiciſſe fideliter artes
Emollit mores, nec ſinit eſſe feros.

OVID. Pont.

amongst these we may enumerate the Chartah Bhade Shattah of Bramah, the Chartah Bhade and Aughtorrah Bhade Shastah, the Vedam*, the sacred books of the Persees, written in the ancient Persian dialects, called the Zend, and Pehlvi†; the Koran of Mahomet; the sacred books of the Mendæans or Sabàites, at or near Bassora in the Persian gulph; the voluminous sacred books of the Lamas in Nexpal and Thibet, called Khangiur, and it's mystic part, termed Riute; the sacred books of the priests in Pegu and Siam, and others of similar tendency. These would be still more valuable, could English, Latin, or French translations be added.

4. Descriptions of the manners, customs, feasts, and religious ceremonies of the respective nations; the architecture both exterior and interior of their temples, religious, public, and private buildings; the figures, names, genealogies and ranks of their divinities and idols; their sacred and domestic utensils, the casts and ranks of people, the learning and religious tenets of each nation, to all which explanatory drawings would be required. What nations use circumcision, and what are the advantages derived from such a custom, or disadvantages from the omission of it? Is circumcision ever extended to the females, and in what manner is it employed?

* In the peninsula on this side the Ganges, the sacred books of the Bramins are contained in the Vedam, copies of which, in the original Sanscrit character, would be very valuable.

† The Pehlvi is a more modern dialect.

5. The

5. The tranflations of the Bible in different languages; and the facred books of Chriftians of various denominations; as the Georgian, Armenian, Perfian, Æthiopian, Coptic, Arabic, and Syriac, efpecially among the Chriftians on the Malabar coaft, and in the ifle of Socatora*.

6. The hiftory and fucceffion of princes, the origin and migration of nations; the government and political conftitution of each country; the caufes of the increafe or decay of power.

7. A relation of the private and domeftic life of the people; the cuftoms obferved at the birth of children; the marriages, fepulchral rites, and any other circumftances characterifing each nation.

8. An account of the aftronomy and chronology of different nations, whether they obferve the fyftem of the feven days of the week, the names for thefe days, with their fignification. The number, names, and fignifications of their months; the number of thefe in a year; whether they are ufed to conciliate the moon's and fun's motion by any intercalation, or a certain cyclus of years; the names for particular ftars and conftellations in the zodiac, with their fignifications; the diftinctions of the other ftars from the planets, with the length of their revolutions.

The ftate in which the art of drawing, carv-

* The Neftorian Chriftians formerly had a fettlement among the Indians on the Malabar coaft, and were there very much refpected. Are fome of thefe ftill exifting? Have they any ancient Syriac books?

ing, and engraving in ſtone has been, or now is. Specimens, drawings, or collections of old inſcriptions, engravings, ſeals, gems, ſtatues, carvings, baſſo and alto relievos, and the places where each of the above monuments are found, the ſize, ſubſtance whereon it is worked, &c. the ancient and current coin, with the exact valuation.

SECT.

SECT. II.

Commerce, Manufactures, Arts, Trade, &c.

JOVE has the realms of earth in vain
Divided by th' inhabitable main:
If ſhips profane, with fearleſs pride,
Bound o'er the inviolable tide*.

FRANCIS.

DESCRIPTIONS and drawings of the looms, tools, machines, &c. employed in manufactures, particularly if ſimple, ingenious, and gaining time or ſtrongly increaſing power, might prove highly beneficial.

2. An account of the planting, gardening, and agriculture of each country; the manure uſed, the time and labour employed in each branch of buſineſs; the price of labour, the implements of agriculture; the kinds and quantities of corn ſown in an acre; the quantities reaped in different ſoils; the proportion of vineyards or paſture lands to arable, and the number of people in a ſquare mile of paſture, arable lands, vineyards, or any other kind of plantation.

3. The kinds of pigments, ſtains, and dying materials known and uſed, particularly in China;

* Nequicquam Deus abſcidit
Prudens oceano diſſociabile
Terras, ſi tamen impiæ
Non tangenda rates tranſiliunt vada.

HOR. l. 1. Od. 3.

are

are they mineral or vegetable? the manner of preparing and applying them, with the advantages and disadvantages of each sort compared with ours; particularly the materials, machines and methods employed by the Indians for dying, staining, and printing their chintzes, calicoes, &c.

4. The wood and timber used for ship building; the form and construction of the ships; the wood employed for masts; the succedanea for oakum, ropes, cables, sails, pullies, &c. with the comparative advantages and disadvantages.

5. The means devised for catching quadrupeds birds, fishes, shells, &c. either for food, or to prevent the increase of such as are noxious to the people or their plantations; are any animals made tame and employed to catch others, or are any methods used for killing or inebriating them?

6. The materials of clothing; if animal skins, the manner of dressing them; if the hair of animals, or the threads of certain insects, the method of spinning, twisting and weaving such substances; if vegetables, how are they cultivated, dressed, spun, and manufactured? The cut and make of the dresses in general, with the advantages and disadvantages of each particular part.

7. The various objects of commerce in general, the growth and manufacture of each article, with the names by which it is known, and it's uses, when designed only for inland trade; the price of labour, and the number of people employed in each department.

8. It is a common opinion, that large quantities of remnants and rags of all kinds of scarlet cloth,

cloth, are yearly carried from England to China, and that the Chinese extract from them their fine red pigments. If this be true, what methods are employed to extract the colour?

9. It has been observed that analogous substances are most proper for dying homogenous bodies*; thus animal substances are best for dying wool and silk, because wool and silk are animal substances. A blue dye, made of woad (Isatis of Linnæus) is found to be full of insects. Is it not the same case with indigo? Are not all the lasting dyes made from animal substances, or of such as contain numerous insects?

10. The manner in which the best indigo is manufactured in the interior parts of Indostan, and the plant from which it is made. Is it from the Indigofera or the Anil? Are there any rules to ascertain when the plant has soaked sufficiently, and how long it ought to be beaten?

11. Is there any linen made of flax or hemp, or what other substances are spun and wove in India besides cotton? What use is made of the yellow or brown cotton taken from the Bombax? Is it manufactured for apparel, and appropriated for a certain order of men, as priests or Bramins?

12. Descriptions and drawings of the instruments and machines employed by the Chinese and Indians to clean the cotton from the seeds.

13. Is only European zaffer from cobalt used by the Chinese for painting their porcelain blue, or have they some of their own? If they have,

* Vid. Histoire de l'Academie, an. 1768. art. 11.

what

what name is it known by, and how is it manufactured? It is probably finer than ours, from the richness of the old China figures.

14. The preparation of the pickle or catchup called Soya; is it made from the Dolichos Soya Linn.? Is salt, wheat or barley added, and in what proportions?

SECT.

SECT. III.

Metereological Observations, Food, Way of Living, Animal Œconomy in general, &c.

For every man to native custom prone,
Conforms and models life to that alone.

GOLDSMITH.

IT is always satisfactory to have regular meteorological accounts by the assistance of a good barometer and thermometer; and to observe at the same time the quarter the wind blows from, and it's degree or violence; the quantity of rain and snow by inches; the size of hailstones; the appearance of aerial phenomena, as aurora borealis, or northern lights, fiery globes, halos or bright circles round the sun and moon; with the effects likewise of thunder storms, lightning, &c.

2. The traveller should also remark the succession of seasons, and the various fruits and productions of each country; the times of sowing or planting, as well as of harvest, or of reaping the grain, &c. the budding or flowering of trees, or shrubs. The food of the inhabitants, and the preparation of it previous to it's use.

3. Some account might be collected of the general prevailing diseases in different seasons, and the causes producing the same, or the remedies employed for curing them, and the methods in which remedies are administered.

4. Are any diseases caused by the effluvia of certain trees or plants, or is the touching, handling or cutting of trees or plants ever suddenly prejudicial to health? Does the effluvia from spice trees or the frankincense tree prove deleterious?

5. What are the effects or symptoms which arise from the bite of poisonous snakes, or any reptile or insect? Has the bite of a snake ever been found to have a salutary effect in curing a certain previous disease, or does the bite of one snake ever destroy the effects of another? Are there any counterpoisons or antidotes usual against the bite of snakes? Is musk or any species of the aristolochias a remedy which the snakes avoid and fly from, or do any of these prove lethal to snakes?

6. Is the pedra de cobra used as an antidote against the effects of the bite of any snake, and with what success? How is this remedy procured and applied?

7. The manner of managing domestic animals, whether in health or otherwise. The animals which the natives in any country castrate, and the effects produced by it; as well as the period when such a custom was first introduced.

8. It would deserve remarking where and in what manner polygamy is introduced, whether perpetual or temporary, and the effects of it upon the manners, religious or civil customs, population, &c. of such countries. Does polygamy prevent some men from procuring wives, or are the women brought from neighbouring nations? Is it

any

any where cuſtomary for one woman to be married to ſeveral huſbands during the life of the firſt?

9. Accurate calculations of births or burials in provinces or towns, and the proportions of males to females, would prove very valuable.

10. Have any buildings or ſhips, furniſhed with electrical rods after Dr. Franklin's method, ever received any injury from lightning?

11. Is the venereal diſeaſe cured without mercury? and if ſo, by what remedies?

12. What diſeaſes attack the workmen employed in different kinds of manufactories?

SECT. IV.

Zoology.

Non ad unam natura formam opus suum præstat; sed in ipsa varietate se jactat.

SENECA Quæst.

1. IT would greatly tend to improve our knowledge in this department of natural history, were the following remarks respecting quadrupeds to be carefully made; viz. the general times of coupling and of gestation; how many young are brought forth at a time, and how often during one season; at what period of life they become prolific or barren: where their principal resort and dens are; whether the males assist the dams in providing food for the young of the carnivorous tribe; how long these are under the protection of the old ones; and what age each species attains?

2. It might be inquired whether any person hath ever seen elephants in copulation, which has been hitherto denied; it is said that if the wild elephants perceive any body, they immediately begin to rave, and cease not till the curiosity of that person has been rewarded with death; and though the Indian princes have kept great numbers of tame elephants of both sexes, they never could procure a breed from them. What differences are there betwixt the African and Indian

Indian elephants? Is the ſtructure of the grinding teeth equal or flat in all, or have ſome elevated or pointed crowns, ſimilar to thoſe of carnivorous animals? Do elephants ever ſhed their tuſks or teeth, or are they permanent?

3. How many ſpecies of tigers, ſo called, or more properly of leopards, panthers, ounces, &c. are there in India, and what are the ſtated and perpetual marks for diſtinguiſhing each, in different periods of life? What other animals of the feline kind are found in India, with the ſpecific characters of each?

4. Does the ſhakal, or jackal, (canis aureus Lin.) bear any reſemblance to an animal commonly called the croſs-fox (Pennant's Syn. Quadr.) What animal do the Arabian writers call bunat-el-auvi?

5. Are armadilloes (daſypodes Lin.) in Aſia or Africa, in what parts, and what are their characteriſtics?

6. When the various ſpecies of the feathered tribe begin to couple or pair ſhould be noticed; when, where, and of what materials each bird builds it's neſt, with the colour, ſize and number of eggs; how long the eggs are in incubation; what the young are fed with, and at what periods they are fledged; with the diſtinctions between male and female birds in different ſeaſons and ages.

7. The migration of birds ſhould not be diſregarded; but their merely diſappearing in one part of the country is not properly a migration, for we frequently find that birds ſhift their place of abode,

at

at certain ſeaſons, on account of ſome palatable food, which may be more plentiful in one part than in another; the croſſing wide ſeas or extenſive continents is underſtood. If any bird be found out at ſea, the ſpecies of bird, the direction of it's flight, the diſtance from land, and the latitude and longitude ſhould be noted.

———In figure wedge their way,
Intelligent of ſeaſons; and ſet forth
Their airy caravan, high over ſeas
Flying, and over lands; with mutual wing
Eaſing their flight: ſo ſteers the prudent crane
Her annual voyage, borne on winds: the air
Floats as they paſs, fann'd with unnumber'd plumes*,

In northern climates it would be uſeful to obſerve when ſwallows are firſt ſeen, and when they diſappear; and likewiſe in what climates they have been found, whether in a torpid or active ſtate, with the ſpecies and peculiar characters.

8. What birds or animals are allowed a privilege or immunity from being injured or killed? and what may be the reaſons for the ſame, or the advantages derived therefrom?

9. Some birds of prey are employed in the eaſt by the grandees in hawking and hunting. What means are uſed for teaching ſuch birds, and what are their differences, ſize, figure, plumage, names, ſpecific characters, &c.?

10. What are the character, plumage, food, and œconomy of the Indian ravens (buceros Lin.)

11. Are there any humming birds (trochilus

* MILTON.

Lin.)

Lin.) in the Indies or China with a filiform long tongue, confiſting of two ſemi-cylinders?

12. What are the peculiar diſtinctions of the true wild peacock? Are there any white peacocks in India, and of a ſeparate ſpecies? Is the change in plumage obvious in wild peacocks, or is this the reſult of domeſtication? Do white peacocks breed with grey and green ones, and what is the colour of the young breed?

13. Is there to be found in the Indian ſeas a jaculator fiſh, ſciæna jaculatrix*, different from the chætodon roſtratus, Linn.? or is the faculty of ſhooting at inſects with a drop of water peculiar to theſe fiſh, or common to any others?

14. Has the rajah torpedo, Linn. or the cramp or numb-fiſh, the ſame electric qualities as the gymnotus electricus?

15. As there is yet wanted a good figure of the ſea cow (trichechus manatus Lin.) it would be deſirable to procure a good drawing of it while alive, to have it diſſected, and to obſerve wherein it coincides with, or differs from other animals nearly related to it, as the ſeal, dolphin, &c.

16. The ſeaſons ſhould be noticed, when different ſpecies of fiſh ſpawn, and the rivers, bays, ſhoals, or ſands they reſort to for that purpoſe; what age they attain before they ſpawn, the food they eat, the age they live to as accurately as poſſible, the ſize they acquire, and the latitude wherein they are generally found; the method of catch-

* Philoſophical Tranſactions, Vol. 54. and Vol. 56. t. 8. p. 186.

ing

ing them, and to what ufes alfo they are applied when caught; and whether they are efteemed wholefome food or the contrary?

17 Which fpecies of moth is it, the caterpillar of which in China affords that ftrong grey kind of filk, and how is it manufactured or wove? How are thefe filkworms or caterpillars preferved, fed, and managed? The introduction of fuch a new filk into England would be a ufeful acquifition, and redeem entomology from the cenfure it is now branded with, of being a mere curiofity void of any real utility.

18. Are there any rattle-fnakes in the Indies or China, or in any part of the world befides America?

19. The various bundles of fwimming fea weeds ought not to be neglected; for befides the different kinds of fuci they confift of, they generally contain fmall crabs, fhrimps, or other fubmarine infects or worms, fuch as the onifci, monoculi, fometimes fhells and efcharæ, fertulariæ, and other corallines.

20. The various animals inhabiting fhells deferve inveftigation, as there are probably new genera yet to be difcovered: it is not fully afcertained whether the inhabitant of the paper-nautilus be an animal really belonging to that fhell, or only ufes it in the fame manner as fome crabs do other fhells; neither is it known whether or no the paper-nautilus animal, or the fepia, increafes or enlarges it's fhell. Indeed the generation, and general œconomy of all the mollufca clafs are very imperfectly known.

21. Experi-

21. Experiments might be tried to ascertain whether pearls are not such concretions in the shells wherein they are found, as the crabstones (lapides seu oculi cancrorum) in the shells of the river cray-fish, which collect alkaline or calcareous materials from the food of the animal for the formation of the new shell, as do the mya and mytilus margaritifer a new layer for the increase of their shells. Do the shells containing the pearls gradually increase through the year, or at a certain season only?

22. It should be carefully remarked which of the shells, especially of the murex and turbo kinds, afford a red or purple juice fit for staining or dying like the purple of the ancients; whether the dye be permanent, and the method of preparing and applying it by some Indian nations. D'Ulloa, in his South-American Voyage, Vol. I. page 182, speaks of one of these shells found near Panama on the Darien Isthmus; and Janus Plancus, or Giovanni Bianchi, in his work on shells, mentions the turbo scalaris Linn. or wendeltrap, as affording the violet purple of the Romans.

23. Descriptions and drawings of the animals or polypes inhabiting the various corals, corallines, sponges, echini, or sea-eggs, asteriæ or star fish, sepiæ or cuttle fish, holothuriæ or sea-nettles, and all the various mollusca, and the polypes of the tubipores, madrepores, millepores, Cellepores, &c. deserve the attention of a natural historian, as but few of them are well described or known.

SECT. V.

Botany.

Itaque iſta quoque naturæ rerum contemplatio, quamvis non faciat medicum, aptiorem tamen medicinæ reddit.

CELSUS.

1. THE ſeeds of almoſt all the Indian plants are worth collecting, they may be preſerved in the manner deſcribed in ſect III.

2. What trees bear the myrobalans, a drug formerly much in requeſt, but at preſent in little eſteem as a medicine, but which might probably afford ſome uſe in dying? What ſpecies of trees bear the myrobalans bellirica, chebula, citrina, and indica or nigra; whether a kind of phyllanthus as the myrobalanus emblica? The uſe alſo of the ſame in India?

3. Is the aloe-wood or eagle-wood, the calambac, and the agallochus, the ſame or different? The place from whence procured, with the generic and ſpecific characters?

4. Is the orchel or rochel (lichen rocella, Lin.) which is found in Madeira, and uſed to dye red, a kind of lichen or a zoophyte? Is the ſteeping it in putrid urine ſufficient to prepare it?

5. Where does the lignum rhodium grow? Is it a kind of nyctanthes or Indian jaſmine?

6. Of what genus and ſpecies is the tea-wood, of which tea-cheſts are made?

7. What

7 What is the wood tek on the Malabar coaſt, of which the Indian ſhips are built? Is it a fact that it is never attacked by ſea worms (teredo, Lin.)?

8. The various kinds of pulſe, as peaſe, beans, phaſeoli, &c. eſpecially ſuch as are reared at the Cape of Good Hope, and exported to the Indies; the other fruits likewiſe which are cultivated at the Cape, and taken in as refreſhments on board the European ſhips.

9. The different kinds of palm trees, their nature, ſoil, characters, names given by the natives, and the uſes they put them to, or their fruits, leaves, bark, pulp, &c.

10. To what genus and ſpecies does the graſs called Tatack belong? and where does it grow, beſides Madagaſcar, Java, the Malay Iſlands, and the French Iſle de Bourbon? The graſſes in general which thrive in particular countries and climates, with the ſoil and culture, and the kinds of cattle moſt addicted to each.

11. What plant bears the famous Indian nut, which is uſed as a reſtorative, and is immenſely dear, being ſold according to ſome at three thouſand pounds each; the place where it is cultivated, the ſoil it requires, and it's real or imaginary virtues?

12 The Columbo root, called by the Portugueſe Raiz de Mozambique, is a native of the continent of Aſia, but it has been tranſplanted to Columbo in Ceylon, and the Dutch now ſupply all Aſia with it. Is it different from the root of Lopez or

 Lopezia;

Lopezia; and if so, what are the characters of each?

13. A description of the small grains and phaseoli, with which the Indians on the Coromandel coast sow their fields after the rice harvest, with the minutiæ of their cultivation, especially the machines employed for watering the grounds.

14. A tree or plant in Cochin China, called Tsai, on being fermented like Indigo plentifully furnishes a green-coloured flour, which in dying gives a lasting tincture of a fine emerald; it would be therefore worth inquiring after the method of extracting the colour, and the additional substances employed in fixing it, and what stuffs are best fitted to receive the same.

15. Are the stamina of the pterocarpus draco, Lin. which is called Draco arbor siliquosa populi folio, by Commelin, and Lingoum by Rumph. Amb. 2. t. 70. connected or separate?

16. Which genus of plants does the true ebony belong to? Is it an aspalathus?

17. Many varieties of rice are found in India, as the red, with red husks; the little rice. small grained, oblong and transparent; the great long rice, with round grains; the dry rice, which grows best on a dry soil, and requires no watering; and the common rice; are these various kinds of rice different species or varieties only? The culture, characters, and specimens of each, if different, would be necessary.

18. There is an elastic gum, called Borrachio in Portuguese, and Kaoutchuck in the language of the natives near Cayenne in South America, of which

which it is said the Chinese make rings for lascivious purposes, but here used by surgeons for injecting liquids, and by painters for rubbing out black-lead-pencil marks. Is this gum manufactured in India or China, and from what plant, and in what manner, with the different uses it may be applied to? Is the plant an Euphorbia or Apocynum?

19. What plants produce gum myrrh and gum ammoniac, and how are they collected?

20. What plant affords the gummi rubrum astringens from Gambia?

SECT.

SECT. VI.

Mineralogy.

And join both profit and delight in one

CREECH.

1. THE manner of working mines, and the methods employed in getting, breaking, and extracting the ores; the tools and machines employed for each of these purposes, are subjects worthy of inquiry. Is gunpowder ever used to blast the stones or ores? The manner in which the ore or metal is found under ground, whether in perpendicular veins, or in vicinity to them; in horizontal flat strata, loose pieces, or in solid continued bodies; in what kind of stone, and at what depth; the means of carrying off the water when present in the mines. The vapours found in them, whether mephitic and noxious, or inflammable when fire or light comes in contact with them.

2. The manner in which white copper, resembling silver, is manufactured, and the various processes whereby it is done.

3. The operations used in extracting the metals from the respective ores, with specimens and names of the ores, and the places where they are

* Simul et jucunda et idonea dicere vitæ.

HOR.

procured;

procured; the products yielded from the ores by fusion; the fluxes added to promote fusion, or the substances to prevent volatilization, and whatever is subservient towards refining of metals or reguluses; with the structures and materials of the ovens, the fuel and quantities of it employed, the time for each operation, and the preparatory cautions, including the picking, pounding, washing, sifting, and ustulating of the ores; and drawings of the various machines and tools used for each purpose.

4. The places from whence the various gems or precious stones are procured, with their prices on the spot; the ground and strata wherein they are found, and the figure or form of each kind before being cut, whether determinate and general, or accidental.

5. The manner of manufacturing those immense quantities of saltpetre annually exported from the East Indies; the soil employed for the lixivium, and the manner of preparing the soil. Are any animal or inflammable substances added to it? By what means is lixivium precipitated? Is an alkali used for that purpose, and how is the alkali procured? Is any use made of the remainder of the lixivium after the precipitation of the saltpetre? Is the lixivium boiled, and inspissated by fire for the crystallization, or by the heat of the sun?

6. If borax be artificial, in what manner and from what substances is it made? If native, in what strata and soil does it lie?

7. How far has the knowledge, value, and use of metals extended amongst nations?

SECT.

SECT. VII.

Directions for taking off Impressions or Casts from Medals and Coins.

————Et cætera penè gemelli.
HORAT. Ep. X.

CHIEFLY owing to the coſt required for purchaſing a cabinet of medals it has happened that the ſtudy of them has hitherto been confined, comparatively, to a few individuals. Another principal impediment to the cultivation of an acquaintance with them, has ariſen from the difficulty of underſtanding the inſcriptions, for want of a ſufficient knowledge of languages; on which account, in particular, this ſtudy has been condemned by the illiterate as barren and uſeleſs; but ſuch as are acquainted with the advantages which have already reſulted from thoſe *nummi memoriales*, cannot heſitate a moment to aſſiſt in promoting a more general purſuit of the ſubject*.

While Coloſſian ſtatues, and the hardeſt marbles, with their deepeſt inſcriptions, are deſtroyed by accidents, or by time, and paintings finiſhed with the higheſt colours quickly fade, a medal ſhall ſurvive innumerable accidents, and diſcloſe hiſtorical facts a thouſand years after ſtatues are

* Though the ſtudy of medals does not properly belong to natural hiſtory, this ſhort account of taking impreſſions from them, may prove acceptable to ſome travellers.

crumbled

crumbled away; and when nothing but the names of an Apelles or a Praxiteles remain. Does not a ſingle medal, of which we are in poſſeſſion, give us greater light into hiſtory, than the once famous libraries of Alexandria and Pergamos, which are now no more? From theſe, and many other conſiderations, I would willingly contribute my endeavours to render this ſtudy more general, and conſequently more uſeful. I have tried a variety of methods to enable a young medalliſt to collect a cabinet, which may initiate him into the knowledge of medals and coins at a trifling expence.

The method of taking off plaſter of Paris and ſulphur impreſſions, is known to every body: the firſt is too ſoft to preſerve them from injury, and the brittleneſs of ſulphur is a greater objection.

By forming a coat or layer of thin metal over the plaſter of Paris, it would be a conſiderable defence. Tin is the cheapeſt and moſt convenient metal for the purpoſe, as it is ſufficiently flexible, and at the ſame time very much reſembles ſilver. The tin-foil ſhould be of the ſame kind with that uſed for ſilvering looking-glaſſes. It ſhould be laid over the medal or coin intended to be taken off, and then rubbed either with a bruſh, the point of a ſkewer, or a pin, till it has received perfectly the impreſſion of the medal; the tin-foil ſhould now be pared off round the edge of the medal, till it is brought to the ſame circumference: the medal muſt then be reverſed, and the tin-foil will drop off into a chip-box or mold ready to receive it, the concave ſide of the foil, or that which is laid on the face of the medal,

 being

being uppermoſt; upon this pour plaſter of Paris made in the uſual manner, and when dry, the figure may be taken out of the box or mold, with the tin-foil ſticking on the plaſter, the convex-ſide being now uppermoſt again, in which poſition it is to be kept in the cabinet, after it becomes dry. To have an impreſſion very perfect, the thinneſt tin-foil ſhould be made uſe of*

The impreſſions taken in the foregoing manner almoſt equal ſilver medals in beauty, and are very durable: if the box or mold † be rather larger than the impreſſion of tin-foil, the plaſter, when poured on, runs round it's edges, and forms a kind of white frame, or circular border, round the foil, whence the new-made medal appears the more neat and beautiful. If this tin-foil be gilt with gold-leaf, by means of thin iſinglaſs glue, or boiled linſeed oil, the medal will reſemble gold.

* This method does not in the leaſt injure any medal or coin.

† Chip boxes, uſed by apothecaries, anſwer this purpoſe, and may be eaſily procured. A ſlip of paper wrapped round any circular body with a flat ſurface, is equally convenient.

INDEX.

INDEX.

Antimonial

Bituminous

Chryſalis

Diſeaſes

Fulmi-

Inſects,

Muriatic

Scarabæus,

Vitreous

THE END.